MEDIA AND DIGITAL LITERACIES IN SECONDARY SCHOOL

new literacies
AND DIGITAL EPISTEMOLOGIES

Colin Lankshear and Michele Knobel
General Editors

Vol. 59

The New Literacies and Digital Epistemologies series
is part of the Peter Lang Education list.
Every volume is peer reviewed and meets
the highest quality standards for content and production.

PETER LANG
New York • Washington, D.C./Baltimore • Bern
Frankfurt • Berlin • Brussels • Vienna • Oxford

Reijo Kupiainen

MEDIA AND DIGITAL LITERACIES IN SECONDARY SCHOOL

PETER LANG
New York • Washington, D.C./Baltimore • Bern
Frankfurt • Berlin • Brussels • Vienna • Oxford

Library of Congress Cataloging-in-Publication Data

Kupiainen, Reijo.
Media and digital literacies in secondary school / Reijo Kupiainen.
p. cm. — (New literacies and digital epistemologies; v. 59)
Includes bibliographical references and index.
1. Mass media in education—Finland. 2. Media literacy—Study and teaching (Secondary)—Finland. 3. Digital media—Study and teaching (Secondary)—Finland. I. Title.
LB1043.K855 371.33'309489—dc23 2012026556
ISBN 978-1-4331-1841-8 (hardcover)
ISBN 978-1-4331-1840-1 (paperback)
ISBN 978-1-4539-0943-0 (e-book)
ISSN 1523-9543

Bibliographic information published by **Die Deutsche Nationalbibliothek**.
Die Deutsche Nationalbibliothek lists this publication in the "Deutsche Nationalbibliografie"; detailed bibliographic data is available on the Internet at http://dnb.d-nb.de/.

Author photo by Kiira Kupiainen

The paper in this book meets the guidelines for permanence and durability of the Committee on Production Guidelines for Book Longevity of the Council of Library Resources.

29 Broadway, 18th floor, New York, NY 10006
www.peterlang.com

Printed in the United States of America

CONTENTS

ACKNOWLEDGMENTS

This book has its origin in my postdoctoral research project, *Youth, Literacies and the Changing Media Environment,* funded by the Academy of Finland, 2009–2010. It draws on and develops further some ideas presented earlier in a few conference presentations and articles, but the analysis, synthesis, and development of ideas and arguments in this book are new.

The idea of *tactic school space* and an analysis of the use of mobile phones in schools in Chapter 3 are taken from my article, "Young People, Mobile Phones and Creative Media Practices at School" (Kupiainen, 2011). In Chapter 4, the discussion of online blogging and forum playing is taken from my article, "Young People's Creative Online Practices in the Context of a School Community" (Kupiainen, in press-a). Discussion of school space and students' creative media practices in several book chapters are taken from my article, "Dissolving the School Space. Young People's Media Production In and Outside of School" (Kupiainen, in press-b).

This study could not be possible without the extreme kindness of the students and teachers of the Maple School where I collected my ethnographic data. The participants assisted with practical problems and gave me all the information I needed. Special thanks go to the head principal, the vice principal, and all eighth grade students and their teachers who gave up their time to work with me. In this study, I use the pseudonym "Maple School," even though I had permission to use the actual name of the school. The reason I decided to use this pseudonym is because "Maple School" typically sounds better to the ear of English-speaking readers and all other names in the book are pseudonyms as well.

I would also like to thank especially Susanna Vehmas, who helped me with the survey data, and the Tampere Research Centre for Journalism, Media and Communication at the University of Tampere, where I began my research.

I began to write some preliminary texts while I was a visiting scholar at the London Knowledge Lab at the University of London in 2010. I would like to thank Professor David Buckingham and the London Knowledge Lab for hosting my work during the autumn term of 2010 and for the stimulating intellectual environment they offered.

I owe a debt of gratitude to Michele Knobel and Colin Lankshear, editors of Peter Lang's "New Literacies and Digital Epistemologies" series, and Mary Shabo, for their comments and suggestions for my manuscript.

Finally, I'm deeply grateful to my wife, Tarja Takala, for her support and patience over the past 15 years, and to our children Kiira Kupiainen, Arvi Kupiainen, and Hilla Kupiainen, for their cheerful good company and guidance on the digital youth culture and media practices.

PREFACE

Public discussion in Finland has been concerned about the digital skills of the youth. Even though the mantra, "digital natives," has celebrated the youth as a generation that has grown up with the Internet, social media, and mobile digital devices, teachers have noted that some students lack basic digital skills. For example, some students are not able to attach a document to their e-mail, if the students use e-mail at all. It seems that the digital culture has divided generationally: young people have different needs and uses of digital devices than older generations. A school is a place where these needs, uses, and practices meet. Young people bring their own practices to the school, and the learning environments of teachers and schools offer their own. In this book, I explore these "meetings," what kinds of practices young people bring to the school, and how the school is structured to meet these practices. I review different media literacies and education that enhance students' digital skills in school. As an ethnographical study, this research concentrates on one secondary basic education school in Finland.

The Finnish school system has been celebrated as one of the world's best. Year after year, the Organisation for Economic Co-operation and Development's (OECD) Programme for International Student Assessment (PISA) results for Finnish 15-year-old students have been excellent (OECD, 2006, 2010). Finnish students are at a high level internationally, especially in scientific, mathematical, and reading literacy. Perhaps the most important reasons for these positive results are a purposeful educational policy and the high standards of teachers. Teachers from primary basic education to upper secondary schools have master's degrees and are highly motivated and appreciated by parents and students (Niemi, 2012). Education in Finland is a public service, free of charge. Basic education starts at the age of 7 and lasts 9 years. Class teachers teach in the comprehensive schools in Grades 1 through 6 and subject teachers take the responsibility of teaching school subjects in secondary school in Grades 7 through 9.

Finland is also known as one of the leading countries in digital technology. Digital devices and the Internet are part of everyday life and embedded in daily practices, from communication to entertainment and content creation. Caroline Haythornthwaite and Barry Wellman (2002) wrote that the Internet is starting to be taken for granted, but this "routineness" is not a reason to ignore it. On the contrary, the Internet as social phenomenon needs more differentiated analysis that takes into account it's meaning in our everyday life. The Internet is not exactly THE Internet, a media rather than an environment and a space, which includes different kinds of media literacies in relation to other spaces, like a school. This study focuses on these literacies that are an inseparable part of the everyday life of Finnish youngsters, and spaces, especially the school, where the students practice their literacies.

Media and digital literacies are part of Finnish core curriculum, not as an independent school subject but as a cross-curricular theme. As a cross-curricular theme, media skills and communication put media education practice in all school levels. One main goal of the theme is to enhance 21st-century skills that citizens need in order to use digital services and be responsible and critical consumers and creative content creators. If we think of some changes in the field of media education, especially in recent years, concepts like *production* and *creativity* have come to supplement the terms *critical perspective, analysis,* and *practice* (Christensen & Tufte, 2010). Young people's own creative media production has challenged the traditional perspective of media education, where the focus has been on mass media and media consumption. A traditional school can also be seen as parallel to mass media, as an industrial-era institution with one-directional learning and teaching settings. When entering school, young people may think they have harkened back to the old days and will not be able to use their own strengths and skills in learning that they have been practising outside of the school, in their everyday life.

Several studies have indicated that young people have learned their digital skills, or media and digital literacies, in informal learning contexts, outside of school with their peers. The use of information and communication technology (ICT) in schools seldom supports students in achieving 21st-century skills. However, informal learning settings do not seem to be sufficient either. For example, some studies have found that most upper secondary school students do not have the necessary skills for evaluating the credibility of information (Kiili, Laurinen, & Marttunen, 2008). The Internet is used infrequently and we still do not have good evidence of the so-called digital natives generation that works collaboratively by using the Internet in a creative way. On the contrary, young people still seem to be more customers than producers and are working more alone than with others (Ilomäki, Taalas, & Lakkala, 2012, p. 67).

In this study, my interest is in creative media production of young people in its own right. At least some young people have started to create and publish content on the Internet and find new ways of learning. They also bring their creative practices to the school community in one way or another by making short films, writing fan fiction, taking photographs, or writing blog entries at school. These practices create new learning spaces to the school but not necessarily to the formal learning settings.

My study indicates that the school space is duplicated. It includes two learning environments, formal and informal, at the same time, and these collide and tangle up in special media practices and literacies. In this book, I explore these practices and their relations to the schooling and school spaces that I call unofficial and official school spaces. These exist side by side at the school, making school boundaries more permeable and offering students possibilities to resist traditional one-sided schooling while finding new ways to learn. But teachers are still necessary as "colleagues" who set new learning settings and help students enhance their critical skills and literacy practices. A new culture of learning is a diverse set of student-teacher and student-student relations that arranges school spaces in a new way and makes a bridge out of school to everyday practices of the youth. Today, learning is omnipresent, happening all around us (Thomas & Brown, 2011).

As it is argued, especially in the realm of critical pedagogy, education is a cultural pedagogical practice that takes place everywhere, including in the mass media, popular culture, and other public spheres. Media education in school spaces and outside has had a goal to enhance young people's critical skills and literacy practices, which are necessary to operate in an everyday media environment. I am interested in cultural media education, where the focus is on everyday meaning making where people, especially children and young people, use media and communicate in different sociocultural contexts. I understand media education, like David Buckingham (2003b, p. 4), as "the process of teaching and learning about the media," where "media literacy is the outcome." Here, *media literacy* is a key term, which refers to broad-based skills and competencies that people must have when they try to communicate, understand, and create media texts and meanings in the modern media age.

Media literacy, as an objective, has often been seen as quite an abstract set of skills. There are two basic problems: (a) Literacy is characterized merely as an individual and cognitive skill, and (b) research is oriented from the administration perspective. In these approaches, the basic question is what kinds of literacies and skills are needed in the information society. The starting point is not in children and young people and their own practices, or what kinds of skills they have and use in their everyday life, but in the question of what kinds of skills are needed by governments and economic sectors. In addition, from this perspective, media literacy is normative.

I am more interested in young people's own media practices, what they bring to the school, how these practices change school spaces and teaching and learning, how they are utilized in schooling, and finally how media education with its own goals helps children and young people enhance their media literacy practices. In this book, media literacy is seen as a sociocultural practice, as multiple and situated within social and cultural everyday practices. From this point of view, literacies are more than individual and cognitive skills. Literacies are situated in particular spaces in different ways and therefore studying these spaces and embodied literacy practices helps us understand meaning-making processes of the youth and what challenges these bring to the school education.

· 1 ·

INTRODUCTION

The Maple School

My research was conducted in a publicly funded secondary school, which I call the "Maple School." The school is located at the center of a city in the western part of Finland. In the academic year 2009–2010, the Maple School had almost 400 students and over 40 teachers. The same building also houses an upper secondary school with about 350 students enrolled in it. The main school is an old neoclassical stone building with four floors. As a school building, it is organized along traditional lines, with separate classrooms, large and reverberating corridors, and a rather small school yard. But near the school is a park, where students may visit during breaks and, for example, make sketches for their visual arts class and film their videos.

One year before I started my ethnographic fieldwork there, the school celebrated an anniversary, and teachers and students made a CD-ROM together where they represented the everyday life of the school from the perspective of students, teachers, and other school staff. On this CD, Grades 7 and 8 students told—in English—about their school life in the following way:

My school, which is very colorful, has lots of pupils.

My schooldays are very loud and energetic, which is fun.

My school is old, but the pupils are young.

My school is an old building, where the yard is full of beautiful maples.

I like my teachers. They are good and nice. I like my way to school. I walk or ride. My schooldays are long, but I don't care.

There are lots of different people in my school, which is very cool.

We have a school yard, which should be much bigger, but it's still good.

School is cool, I think that's true. My friends are there.

The most important moments in my schooldays are the breaks when us friends can talk to each other and laugh together.

My school has a large TV, that cost over 1000 €, but no place to sit, which is very stupid.

My favorite places in the school are the canteen and the computer lab, because they both are very cool places.

Many students mentioned that the most important things for students at the school are friends and places like the schoolyard where they can spend free time between classes. Some of them said that sometimes they hate school, but usually it is okay, even though they have long days and school subjects may be boring. Saying this is also a part of the school discourse of teens, but looking at the Health Behaviour in School-aged Children (Currie et al., 2004) study, we find that among European countries, children in Finland dislike school on a large scale. At the same time, however, they report very high life satisfaction in general. This low level of school satisfaction has been under debate in Finland since World Health Organization (WHO) reports (Currie et al., 2008) showed that the trend has been more positive lately; however, this upward trend is not moving fast enough to satisfy school administration and policy makers. Reasons for low school satisfaction in Finnish schools are hard to find. At the same time, teachers are highly motivated, even though the school culture has been under change and the climate of peace in classrooms has decreased. Usually, reasons for low school satisfaction are traced back to Finnish society in general, which has taken a more "hard-line" direction, where market rationality dominates everyday life more and more. Children and young people have more pressures and expectations placed on them by society and the school and not so much time and space just to be and build their identity in peace.

The principal of the Maple School complained that school subjects require much more from students than before, and society puts more pressure on schools. Teaching is not the only role that teachers now have. At the same time, they have roles of police, priest, psychologist, and even parent during school days.

The Maple School is not special when it comes to school satisfaction, but it has some emphasis on art teaching. Visual arts have compulsory status in Finnish schools' curriculum, but after Grade 7 (i.e., in Grades 8 and 9), it is an optional course. However, in the Maple School, some of the required classes are special visual arts classes, which means that they have visual arts lessons also in Grades 8 and 9. In Grade 9, students also make a special productive thesis of visual arts. In short, half of the classes have a visual arts emphasis and three classes are combined so that there are both students with and without visual arts.

According to the Finnish Youth Barometer, young people think that school has a major role in promoting creativity. Actually, over half of the respondents ages 15 to 29 reported that there should be more common arts education in school, with only minor differences reported between the opinions of females and males (Myllyniemi, 2009). Finnish students' attitudes toward art subjects (crafts, visual arts, music, and physical education) are clearly positive (Kairavuori & Sintonen, 2012, p. 211). Seija Kairavuori and Sara Sintonen argued that

> It is not only the question of cognitive development in other dimensions, or a break from the demands of academic subjects or pure enjoyment, but also for example in improving cultural and communicative reading, producing and learning, in understanding Finnish traditions and culture and as part of individual and social growth. (2012, p. 211)

The principal of the Maple School pointed out that visual arts emphasis brings some advantages to students' schoolwork:

> My opinion is that it is a clear advantage. And especially for sensitive boys it is usually some kind of salvation, when they do not need to be ice hockey players that boys usually admire, so it is a really important thing for them. And at the school level, students in the visual arts classes are quite liberal; they accept differences and are devoted. Visual arts classes have, for example, more positive attitudes towards studying, for some reason, even though the students are selected only on the grounds of their skills in visual arts. (Interview with the principal, male)

The principal indicated that visual arts students are especially interested in doing things actively, having different projects, and working together. For these reasons, some other students were also moved to the visual arts classes, where they had more possibilities for expressing themselves and being recognized positively by others. Visual arts students were deemed more tolerant of others, too. The visual arts subject area at the school certainly gives more opportunities for students to find their personal strengths and skills as learners, as we will see later.

The Maple school has more girls than boys. In my survey at the school (N = 305), 68% of the respondents were girls. Especially in visual arts classes, girls were the majority. Therefore, the comparison between the sexes in my study is rather difficult to do.

My ethnographic field work in the school was conducted between August 2009 and June 2010. I observed in the school from 3 to 10 hours a week, mainly mother tongue and visual arts lessons, because media education has been explicit in these classes and I was especially interested in how media education is conducted at the school. In addition, ICT teaching was my particular focus. The most regularly observed group was a cohort of eighth-grade students, who also had an optional movie course as part of their curriculum.

The report of the assessment of goals of cross-curricular subjects in Finnish schools by the Finnish National Board of Education shows that media education tends to be formally implemented in schools in Grade 8, when students are 14 to 15 years old (Niemi, 2012, p. 25). My experiences at the school support this finding. In Grade 7, students have just started secondary school and are getting accustomed to the new learning environment with its different subject teachers and so on. In Grade 9, they start to seek new study trajectory possibilities for when their basic education is completed at the end of this grade. Therefore, they tend to concentrate more on official academic subjects. But in Grade 8, the students have more "room" to do different kinds of projects. In tandem with this, the mother tongue language textbook used by most schools is more focused on media literacy issues in Grade 8.

The basis of my data includes handwritten field notes taken during observations and at the end of each day. During my ethnographic field work, I also conducted 34 formal, semistructured interviews: 26 with students and 8 with teachers. All interviews and some of the lessons in the school were recorded by a digital audio recorder and transcribed afterward. During my observations, I used video and photo cameras in some cases as well. For example, I filmed seventh-grade students when they presented their own soap opera improvisations in a mother tongue language class and students making a film in the video course. A photo camera was used mainly in corridors outside of the classrooms. I also photographed some of the students' works that were created in different classes. Additionally, I gathered students' media analyses, essays, and the video films that they themselves had created. My ethnography was preceded by a quantitative survey (N = 305) on media use among the participants.

My ethnographic field work was not limited to the school. I also observed some students in their homes when they, for example, played video games or tried to create a new Facebook profile. A part, albeit an important part, of my ethnography was *netnography* (Kozinets, 2010) or virtual ethnography (Hine, 2000), a form of ethnographic research on the Internet and on social networking practices. For example, I followed some of the students' weblogs, YouTube channels, social networking profiles on the IRC-gallery,[1] and deviantART (http://www.deviantart.com) pages.

School Ethnography

At the beginning of my study, I had plans to explore the various meanings of media among young people, how they embrace media, and how media is integrated into their everyday lives and schoolwork. I generated the following research questions:

- What media technologies and contents do young people use in home and school?
- What kind of literacy practices do young people engage in within home and school contexts?
- How are home, school, and media linked together in the media use of young people?

During my observations, however, I realized that it is not necessary to explore home and school practices as separate entities. It was more important to understand what kinds of uses and practices young people have in general and how these are visible in the school context. School, of course, is a meaningful place for these youngsters and is not totally separated from other parts of everyday life, such as home, free time, and online communities. As students wrote in the school's anniversary CD-ROM texts, friends and some school spaces have especially important meanings for them. Some students also mentioned the computer lab as an important part of their school lives, referring here to their online lives. Even though almost all Finnish households have a computer and access to the Internet, for some students, the school may be the only place where they can use the computer and the Internet at all or without mediation by adults. My survey suggested that only 1% of students did not use the Internet at all or less often than once a month, and 72% of the students used it every day. The EU Kids Online survey suggested that 79% of Finnish 9- to 16-year-olds use the Internet every day or almost every day (Livingstone, Haddon, Görzig, & Ólafsson, 2011a, p. 25), and 84% go online at school. Thus, online life is a substantial part of young people's everyday and school lives in Finland. Therefore, the school ethnography is also intertwined with observing online practices and everyday life in general.

Ethnography has been a common methodology both in media studies and education. Media ethnography has especially been interested in media use and audiences of different kinds of media. For example, in the late 1980s and the early

1 IRC-gallery is one of the most popular Finnish social networking sites among youth. In the last 2 years, it has lost its popularity to Facebook, but it still has 450,000 registered users and young people submit about 1 million comments daily.

1990s, researchers in developed countries focused on television reception in households. The subject of television and children has also been an interesting area of focus among ethnographers. For example, Dafna Lemish and Mapel R. Rice (1986) observed young children (6.5–29.5 months old) in order to study how television affects language acquisition. In the same year, Australian researchers Bob Hodge and David Tripp (1986) published a well-known book, *Children and Television,* in which they combined semiology, cognitive psychology, and ethnography. Hodge and Tripp's study also examined the mutual relationship between home and school and demonstrated how television is within both of these.

The history of educational ethnography is longer and wider than the history of media ethnography. Tuula Gordon, Janet Holland, and Elina Lahelma (2001, p. 189) wrote that ethnographic research in educational settings "has a history of more than three decades in a number of countries.... " Predominantly, field work has been conducted in an area of formal education, like schools with students ages 7 to 16. Needless to say, there are also other areas of foci, too: for example, kindergartens and adult education contexts.

In relation to school cultures, ethnographers' interests have been mostly in processes of negotiation, sex and gender, social class, and different classroom procedures. School is often seen as a separate entity, a microcosm that is adapted from the rest of society. It is common to speak about a "school culture," which is seen as distinct from, for example, a "home culture," or at least some discontinuities are seen between the two. In my study, I tried to see school culture as "mixed culture," where school culture and students' everyday culture tangle up. Rather than a one-dimensional cultural place, a school is multicultural in many senses. As Margaret Eisenhart (2001, p. 20) argued, it is worth asking what views of culture can guide the ethnography of groups and sites of permeable boundaries, multiple influences, and dispersed networks. In the new literacy studies, culture is seen as a verb rather than as a noun (Heath & Street, 2008, p. 7). Culture is then dynamic, unbounded, and kaleidoscopic without clear boundaries. Jan Nespor (1997) defined culture in his school ethnography in terms of "funds of knowledge," which are made available in networks of activities and communities that intersect in particular times and spaces. Schools and classrooms then have intersections of multiple cultural resources and different kinds of contexts, some local, some global; it is like a knot in a web of different practices. Popular culture, mass-distributed images, and representational spaces are one part of the funds of knowledge that are present at school.

Eisenhart (2001, p. 22) argued that ethnographers should look beyond a single classroom and school, "upward and outward" to the different cultural forms that are manifested and produced in networks of larger social systems. At the same time, ethnographers should look "downward and inward" to see how these cultural forms become part of particular spaces and individual subjectivities:

> Ethnographies will have to include ways of exploring the connections among sites that together make up arenas of social practice, such as among the households, schools, extracurricular activities, personal relationships, TV shows, video games, and transportation networks that connect up or intersect to form the contemporary contexts of youth activity.... (Eisenhart, 2001, p. 22)

From this perspective, studying media literacy in school is not about conducting traditional educational ethnography where a researcher is interested especially in school culture, teaching practices, or school literacies. Rather, interest is in the knots of a "literacy web," where school literacies and out-of-school literacies collide.

School has a specific kind of space (classroom) and time (lesson), which differs from out-of-school spaces and time, and the use of media devices is somewhat different than in the home. Students also become socialized to classroom practices and schooling in general. This means that they start to think and act in terms of formal school practices. For example, students can say that the Internet does not belong in school, where the most important area of focus should be to anticipate an examination. Regardless of such claims, however, they still bring their own literacy practices to school.

Kevin M. Leander (2009) suggested that ethnographic research offers possibilities for learning about media and digital literacies as social practices. He used the term *connective ethnography* and the alternative terms *social spatial ethnography* and *traveling ethnography* to describe an approach to studying Internet-related practices "as normative social practices and Internet social spaces as complexly connected to other social spaces" (p. 37). This approach tries to avoid traditional dichotomies between online/offline, virtual world/real world, and cyberspace/physical space. The Internet is seen as a social space and "heterogeneous assemblage of practices, texts, and cultures" (p. 60) connected to transformations of everyday life and circulation of texts, things, and people. From this perspective, traditional separation of school and out-of-school literacies can be seen as "forms of identity building, or producing communities, social groups, and institutions" (p. 61). My approach is to use connective ethnography as a guideline and see different literacy practices connected together in different (school) spaces and as a part of learning and identity practices. All the same, it is difficult to avoid dichotomies between different text types, but these have to be seen more as a means to study literacy practices and generate analyses in order to understand these diverse practices.

Ethnography does have some drawbacks, however. Conventional ethnography within media studies is more often interested in media environments and texts. But as Steve Jordan (2003, p. 92) argued, this kind of research often fails to understand how these kinds of cultural processes are embedded within the contemporary accumulation process. Jordan offered a critical ethnography, where the focus is not so much on literacies but on a new cultural economy, social relations of cultural accumulation, and new types of hegemony:

> Such an approach is suggestive of a complex integrative analysis that does not treat culture, ideology, "the text," images, or technology as separate categories, but rather understands them as a part of the historical and material processes of capitalist accumulation that feeds off human labor power. Critical ethnography's strength, therefore, lies in its capacity to produce finegrained ethnographies of how this process organizes and exploits human labor power in the production and consumption of cultural commodities under global capitalism. (p. 96)

This critical approach reminds us about the dangers of celebrating literacy practices without seeing how these are part of broader contexts of cultural pedagogies and often are in the structural control of commercial companies. The Internet, for example, offers a space for self-expression and a publishing platform for creative literacy practices, but it is double-sided; users of the Internet are quietly "peeked at, poked, analyzed, and tagged as they move through the online world" (Turow, 2011, n.p.). Google cofounder Sergey Brin expressed his deep concern about the freedom of the open Internet in an interview with the *Guardian* (Katz, 2012). He mentioned how big companies like Facebook dominate the Internet and rule us and how Internet censorship has been tightened in several countries. For his part, Joseph Turow (2011) argued that marketing companies are performing social profiling and discrimination by customizing our media content on the basis of our behavior on the Internet. This also gives context to literacy practices where highly celebrated participation in content creation may be more pseudo-participation in a minimalistic way. David Buckingham (2010) pointed out the following:

> It would be wrong to assume that participation is always a good thing, or that it is necessarily democratic, counter-cultural or liberating. Creative production can be a powerful means of learning—whether it involves remixing of various kinds, appropriating and adapting existing texts, or creating wholly new ones, or simply exploiting the potential for networked communication. However, all this needs critical reflection....(p. 301)

New youth literacy practices have the potential for more democratic and liberating cultural activities and learning styles, but this should not be taken for granted. Ethnographic study helps us understand and see these practices in different contexts and, in my case, within the school context. But there are a lot of things that an ethnographic study cannot do. For example, it is not an ideal approach for seeing how companies track and segregate people on the Internet and control their literacy practices, because we rarely have access to observable data concerning such control. We also have to keep this limitation in mind in educational contexts.

· 2 ·

STUDYING LITERACY PRACTICES

Media and Digital Literacy

In this book, my objective is to understand the ways in which young people make meanings and live in the world when they use different media technologies and media texts and—especially—when they produce and share media texts and content, as already mentioned. That is why it is important to shift our focus to the social practices of meaning making, rather than to focus on a pure set of technical and cognitive skills (Gustavson, 2008, p. 82). I understand young people's media practices as kinds of creative practices. Leif Gustavson argued the following about creative practices:

> The creative practices of youth, whether skateboarding, gaming, redesigning cars, or Parkour, are part of their everyday lives. They literally carry the practice with them wherever they go. Youth spend hours and days practicing their craft, swapping techniques, and scrutinizing their own practice. These creative practices take youth to all kinds of spaces: friends' houses, conventions, skate parks, studios, concerts, clubs, and stores where they know they will find people and texts that are interesting and will inform their work....Their creative practices are part of who they are and how they understand the world around them. (p. 82)

Gustavson does not mention school as one of these spaces where creative practices are present. If we think that a school is some kind of "knot in a web" of different everyday practices, it is worth asking what kinds of creative practices young people bring to the school and how teachers and formal schools meet these practices. Rebekah Willett (2009, p. 21) explained: "As educators, we need to see what young people are doing, so that we can build their existing practices and knowledge, and identify barriers and enablers to new media literacies." Media and digital literacies are creative practices through which young people construct their identities and act in the digital world as citizens and participants of different offline and online communities. The question is how these practices are present in everyday school life and how they "meet" school spaces and formal learning, as already mentioned.

According to the standard definition, media literacy is "the ability to access, understand and create communications in a variety of contexts" (Ofcom, 2006). This definition is based on the so-called Aspen definition, which had four parts originally: access, analyze, evaluate, and produce (Tyner, 1998, p. 120). From this perspective, media literacy is a rather abstract individualized and cognitive skill that gives individuals the capacity to find information, for example, from different websites and data banks, and use media (access), understand and critically evaluate information in different forms (understanding), and also produce media content for their own purposes using different media forms (creation).

The European Commission also uses the same kind of definition in different contexts. For example, the Study on the Current Trends and Approaches to Media Literacy in Europe includes the following definition:

> Media Literacy may be defined as the ability to access, analyse and evaluate the power of images, sounds and messages which we are now being confronted with on a daily basis and which are an important part of our contemporary culture; as well as to communicate competently using media available, on a personal basis. Media literacy relates to all media, including television and film, radio and recorded music, print media, the Internet and other new digital communication technologies. (European Commission, 2007, p. 13)

This definition has an emphasis on communication and takes a step further into a focus on the production of one's own media texts and contents.

If we look at some definitions of digital literacy, we can see the same kinds of key competences come into play. For example, Paul Gilster (1997, p. 33) defined digital literacy as "the ability to understand and use information in multiple formats from a wide variety of sources when it is presented via computers" and through the Internet. Gilster identified four key aspects of digital literacy: content evaluation, hypertext navigation, searching, and knowledge assembly (the latter refers to the ability to collect and customize knowledge for one's own purposes). Digital literacy differs from media literacy in most cases from the perspective of technology and context: media literacy traditionally has its technological focus on mass media and audiovisual media as well as print media. Digital literacy is more focused on digital environments like the Internet and computing. Of course, they have much in common because of the digitization of the media environment in the last two decades.

As Colin Lankshear and Michele Knobel (2006, p. 21) commented, the debate concerning digital literacies is often associated with fears of a digital divide, between digitally literate and illiterate people. In many governmental strategies and policies, digital literacy plays a double role of combining social and economic dimensions of literacies. The social dimension of literacies is shaped with the economic purpose of contributing to the growth and competitiveness of societies. Competitive and innovative knowledge-based economies in the global markets need good literacy skills of citizens and the workforce. Hence, digital literacy is increasingly incorporated into the process of governance, the labor market, and global competition (Livingstone, 2008, p. 56; Rantala & Suoranta, 2009, pp. 100–101; Rodriques, 2006, pp. 412–413). But at the same time, it includes the themes of democratic participation, active citizenship, lifelong learning, cultural expression, and personal empowerment (Kotilainen & Suoranta, 2007, pp. 4–5).

From the perspective of sociocultural literacy theory, the term *literacies* challenges the dominant emphasis on a single form of *literacy*. There is a plurality of literacies in different literacy events and domains of life. The way to study these literacies is not to set up some general criteria for literacies. Rather, it is to look at the ways in which people engage in everyday literacy practices.

I understand media and digital literacies from this perspective as social practices. With this definition, I refer to the New Literacy Studies and a sociocultural perspective of literacy. That is, "Literacies are bound up with social, institutional and cultural relationships, and can only be understood when they are situated within their social, cultural and historical contexts" (Lankshear & Knobel, 2006, p. 12). James Gee (1991) argued that literacy is linked to different discourses, ways of being in the world, and social identities. Therefore, reading and writing is a different practice, for example, in an online chat room than in a school classroom. Youth culture outside school is filled with multimodal texts in affinity spaces and it is noticed that here lies the basis of difference between school and out-of-school literacies. In general, both have different practices concerning talking, acting, thinking, valuing, writing, and reading. But when we look at different literacy practices in school, we notice that young people bring their own practices with them to the school and such practices ultimately tangle with the school's practices. Media and digital literacies are a matter of not only knowing how to encode and decode texts but also applying this knowledge for specific purposes in different social contexts. Media practices like blogging and writing fan fiction, are examples of these kinds of application of processes and are the focus of this study.

From the perspective of a sociocultural theory of literacies, we are able to speak about "new literacies." In the era of the Internet and "posttypographic" forms of texts, some social practices have changed and new forms of literacy have emerged. As Lankshear and Knobel (2006, p. 25) argued, many "of these new and changing social practices involve new and changing ways of producing, distributing, exchanging and receiving texts by *electronic* means."

Digital and social media are examples of new electronic means. In the term *social media,* the emphasis is on the word *social.* When people are using social media they are no longer only audiences that simply receive messages. Rather, they also create content (e.g., by writing weblogs or Wikipedia entries) and share meanings and generate discussions (e.g., for example, on Facebook or Twitter). This means that media and digital literacies are more participatory, collaborative, and distributed in nature than traditional individual and author-centered conventional literacies (Lankshear & Knobel, 2007, p. 9).

But at the same time, we have to remember the critical aspects of media and digital literacies. Sharing and participation as new modes of literacies are not enough; we also need people to be critical participants "and to develop a broader understanding of the economic, social and cultural dimensions of media," as David Buckingham (2010, p. 299) argued. It is not an overstatement to say that the technology itself does not automatically produce social change and new, valued forms of media and digital literacy. Digital technology offers some possibilities to engage in an active and participatory way, but it requires critical thinking and analytic skills as well as active participation and valued practices of message and content production.

Domains and Sites of Literacies

In this study, I was especially interested in different places and spaces of media literacy. Different spaces require different literacy practices. As argued by Barton (2007), there is more than one way to read and write. Reading a newspaper, surfing the Internet, and watching television involve different literacy practices, and these often get mixed up in everyday practices. Digital natives are especially purported to have the skills to use a variety of media devices at the same time and to read different types of texts. This is called *multiliteracies.*

Different literacies are associated with different domains of life such as school, home, and work (Barton, 2007). Different places enable different roles and language use. But at the same time, for example, home is a typical place for a variety of literacy practices. Students do their schoolwork at home and their parents complete work-related tasks in home spaces. Different practices leak from one domain to another and become mixed. As argued in the previous chapter, as a physical location, home is a site for different types of literacies. This applies to the school as well. School includes and requires different kinds of literacy practices, although school literacy does play a dominant role, especially in formal learning.

Additionally, literacies are not neutral. They fight for place and time. This is easy to see at school where students' own media practices usually do not have room in class. Classes are dedicated instead to school literacies. For example, using mobile phones or MP3 players is generally forbidden in Finnish schools. If students have extra time at the end of the class after completing their assignments, they might obtain permission to use the computer and go on the Internet. It is a type of prize after hard work but not a recognized or valued part of schooling and formal learning.

Institutional and Vernacular Literacies

A school is an institutional place, a site of literacy that is an example of the dominant or institutional literacy. Mary Hamilton (2000) defined institutional and dominant literacies in the following way:

> Dominant literacies...are associated with formal organisations, such as those of the school, the church, the work-place, the legal system, commerce, medical and welfare bureaucracies. They are part of the specialised discourses of bounded communities of practice, and are standardised and defined in terms of the formal purposes of the institution, rather than in terms of the multiple and shifting purposes of individual citizens and their communities. In institutional literacies there are professional experts and teachers through whom access to knowledge is controlled. To the extent that we can group these institutional literacies together, they are given high value, legally and culturally. Institutional literacies are powerful in proportion to the power of the institution that shapes them. (p. 4)

The institutional framework is usually an important part of everyday practices around literacy. Schools, for example, support a certain understanding of literacy and have certain purposes and control over people's literacy practices. On the contrary, vernacular literacies are learned outside school, "on the streets" (Lankshear & Knobel, 2003, p. 31). Vernacular literacies are associated with everyday life, outside the domains of power. They are learned informally and are not regulated by the formal rules and procedures of any social institution. Hamilton (2000) wrote that they are

> more common in private spheres than in public spheres. Often they are humorous, playful, disrespectful, sometimes deliberately oppositional. When questioned about them, people did not always regard them as real reading or real writing. Some vernacular literacies are deliberately hidden: these include those which are personal and private, where reading or writing are ways of being alone and private, ways of creating personal space. There are also secret notes and letters of love, abuse, criticism and subversion, comics, scurrilous jokes, horoscopes, fanzines, pornography—some but not all of which will be revealed to the researcher's gaze. (p. 7)

If we look at young people's media practices, we see them learning literacies out of school, within peer-to-peer relationships. Some of these literacies are private and hidden, but some are present at the school where students bring them. Lankshear and Knobel (2007, p. 19) gave us many different examples of vernacular literacies that are part of young people's everyday lives, like blogging, writing fan fiction, or collaborating on Wikipedia entries. With these practices, it is important to feel connected to other people and to celebrate, for example, a fandom: "to participate in an affinity, to make shared meanings, to brighten the day, share a laugh, share one's passion for a product or a character, and so on" (p. 13). Vernacular literacies activate people to get involved in some way with groups, communities, or spaces, where they read and write together. In this way, literacy is more a communal resource than an individual skill.

My focus in this book are the literary practices of blogging, writing fan fiction, making videos, and so forth. They are out-of-school literacies, but they are connected to school in a special way. Even though the school is a realm of the book and book-related literacies, it is also a space for young people's everyday lives and practices as "ways of creating personal space" (Hamilton, 2000, p. 7).

I also focus on school literacies and media practices in curricular learning contexts at school. School is a place where students are able to learn about the media and become more critical consumers and active media producers and where they can obtain some conceptual understanding about the media world and media environments. This is an important part of growing toward active and critical citizenship.

Media Education in Finnish Schools

As mentioned before, media literacy is a cross-curricular theme in Finnish national core curriculum. This theme is integrated into many school subjects, although it is found mainly in visual arts and mother tongue classes. This theme in the secondary basic education curriculum is called "media skills and communication." In the core school curriculum, the objectives of media skills and communication comprise the following:

> The goal of education is to develop expression and interaction skills, to advance the understanding of the status and significance of media and to develop an operating knowledge of the media. In communication skills, the emphasis is on committed, interactive and communal communications. Media skills are practiced both as recipient and originator of messages. (Lundvall, 2009, p. 12)

As we can see, in the national core curriculum, the term *media skills* is used instead of media literacy. The term *media literacy* is difficult in Finnish language, without a distinction between the terms *literacy* and *reading skill*. Literacy is understood in the Finnish language traditionally as the ability to read only literal texts, and therefore many teachers do not accept that the word "literacy" can be used when we are talking about media.

Using the terms *media literacy, visual literacy, digital literacy,* or *information literacy,* therefore, is still difficult within the school context in Finland. However, media literacy is an international term that describes an ability to be literate. To be literate means to participate in society, discuss and share messages in public spaces and with a critical attitude, and use the media for individual purposes. The Finnish National Board of Education has also used the term *extended literacies* to refer to the wide understanding of various texts from literal to visual, and so forth.

As a cross-curricular theme, media skills and communication have become a part of national Finnish school culture. The responsibility to teach media literacy from this perspective is divided among teachers of all school subjects, but, as I mentioned earlier, the main responsibility typically lies with visual arts and mother tongue teachers. However, the evaluation of the cross-curricular theme of media skills and communication has shown that teaching and learning media and media literacy have an obvious impact on students' media knowledge and skills (Kotilainen & Kupiainen, 2012). This evaluation also shows that students learn media at home and in out-of-school spaces as well.

Teachers of ninth-grade students in Finland have reported a strong interest in teaching media skills and communication: 91% have said that it is an important or extremely important theme. Among all seven cross-curricular themes, media skills and communication are third in regard to teachers' interest. Furthermore, 93% of teachers thought that the theme is necessary or highly necessary (Niemi, 2012). Accordingly, teachers are very motivated to teach media literacy, even though it is not an individual academic subject.

Nonetheless, the voluntary status of media literacy has brought with it some significant problems for equal attention across schools and even across classes. The implementation and practices of media education depend on the attention and activity of individual teachers and schools. Therefore, there are huge disparities across schools and districts with respect to media education. Especially in rural areas, media education remains underdeveloped and is little practiced within basic school subjects and contents. Furthermore, there are inequalities among teacher training programs with respect to preparing teachers to be effective media educators. Although teacher training in Finland is of high quality, it is still possible to graduate without having completed a single course in media education. However, in some universities, media education is a core theme within teacher preparation programs. The situation thus varies considerably in teacher training programs across different universities.

This may be one reason why there seems to be some kind of generational gap between students and teachers in schools. An extensive Finnish study found that teachers and their students have very different literacy practices: "There is a growing gap between practices in school and the way in which students use various media in their free time for informal learning and social existence" (Taalas, Tarnanen, Kauppinen, & Pöyhönen, 2008, p. 240). Information for this survey was collected from about 1,700 students and 740 mother tongue and foreign language teachers in 102 Finnish secondary schools. The survey showed that digital materials were used quite scarcely among teachers in classrooms: for example, 66% of mother tongue teachers had never used or worked with web-based materials in a classroom. Also, over half of the foreign language teachers reported that students in their classroom were never engaged with video or audio recordings, which is hard to believe because these learning materials have been used for decades in Finnish schools. For 98% of foreign language teachers, the textbook was the primary teaching tool (Taalas et al., 2008, pp. 248–249).

Students' media ecology seemed to be quite different. For them, the new digital media was the most important way to spend time: only 6% of the students reported not using online resources in their free time (Taalas et al., 2008, p. 251).

The EU Kids Online survey showed that children and young people from 9 to 16 years of age across Europe use the Internet mainly for schoolwork, but in Finland, the percentage of Internet use for schoolwork was far below the top countries. In fact, Finland was among the bottom 5 of 25 countries (Livingstone et al., 2011a; Ólafsson, 2011). This suggests that the new media do not have a great part in Finnish schooling, and the traditional book is still a crucial learning tool for teachers and students. The public debate has focused on old-fashioned pedagogy that does not fit the media ecology of the 21st century. The student-computer-Internet ratio in Finnish schools is rather good, but teachers use ICT for teaching rather poorly (Kankaanranta & Puhakka, 2008). And if teachers are not familiar with new media and ICT, how are they capable of teaching media and media literacy and understanding the everyday lives of the youth?

But things may change rapidly. In the spring of 2012, the public debate in Finland has referred to many school projects where iPads and other mobile devices have started to be used as part of learning settings. I would also add that although the school differences are not relatively large in Finland, there are differences after all. In some schools, use of ICT and media in teaching and learning as well as media education are routine practices, but in other schools they are not. The gaps between schools seem to have widened rather than shrunk (Vahtivuori-Hänninen & Kynäslahti, 2012, p. 240).

· 3 ·

MEDIA ECOLOGY IN SCHOOL SPACES

Social Spaces

Media literacies are connected to different sites and spaces. For young people, school is one of the most dominant places and spaces where they spend several hours per day. It is also a space for different media use and practices, both for individualized media practices and school-based media practices during classes. But when we are talking about the media, we also use metaphors of space: cyberspace, media environment, public sphere and space, and so forth. "[M]edia have an 'environmental' quality," as Shaun Moores (2012, p. 13) pointed out. Technology and media also reconfigure space and time, restructure geographies, and propose new ones (Papacharissi, 2010, p. 7). It is said that the world is smaller because of the media and the Internet, as communication between distances is easier due to communication networks, mobile phones, and teletechnologies. Zizi Papacharissi (2010) noted the following:

> Much like geographically fixed worlds that humanity has conquered, these technologically enabled new worlds present imagined geographies that individuals hope to reinvent themselves in and tame at the same time. (p. 9)

The Internet especially is seen to be some kind of new world and homestead for new communities. The cultural approach to Internet study sees the Internet as more of a cultural and social space than as technology or simply a cultural artifact:

> To study the Internet as a culture means to regard it as a social space in its own right, and to explore the forms of communication, sociability and identity produced in this social space and how they are sustained by resources available in the online settings. (Mesch & Talmud, 2010, pp. 4–5)

My approach is to see the Internet as a place for being and dwelling and in relation to other spaces, like school space, as argued before. Traditionally, space has been seen as a physical realm where embodied human beings and other three-dimensional objects lay. But everyday space as well as media space is something more than that. Online and off-line spaces are also mixed. As Rodney Jones (2005) wrote, cyberspace "does not create a 'separation' of spaces so much as additional layers of space within which participants can maneuver" (p. 142). As Jones argued, spaces arise from actions. Our actions do not occur "in" or "at" spaces; actions create spaces.

After the spatial turn in social and human sciences in the 1970s and 1980s, space and place have been important subjects for studies in different disciplines, such as anthropology, sociology, human and cultural geography, and spatial humanities. Space is not an easy concept, and as Henri Lefebvre (2007) pointed out in the opening of his major work, *The Production of Space*, it was not long ago when the word "space" had only geometrical meaning: "The idea it evoked was simply that of empty space" (p. 1). According to Lefebvre, to think of space as a geometrical dimension is not enough; we have to take into account lived practices of the space as well as symbolic meanings of the space. Space exists only as inhabited and lived. My approach is to study school space from this point of view, where lived practices like teaching, learning, consuming, using, and making media construct the space. The school is more than a building with different rooms for students and teachers; it is a living space, or better in plural, spaces for different practices and social interactions. Of course, formal learning is one of the most important practices in the school spaces, but there is also room for other kinds of practices, like media practices. In brief, media literacy practices are part of the constitution of the space, and in my case especially, part of the school space, consisting of the relationships among participants.

After the spatial turn, space has been seen as an inseparable part of everyday social life and meaning making. Space still has physical dimensions, but as human beings, we are spatial beings who make room and encounter other beings in social spaces. We live in our homes, walk in the city and on pathways, surf the Internet, and go to school. Spaces are not empty but full of life and different everyday practices.

Doreen Massey (2010, pp. 9–10) gave us important propositions to understand space. First, space is the product of human interrelations. This also means that space does not exist prior to identities and their relations. These are part of spatiality. Space is constituted in different relations, which are understood as embedded everyday practices. At school, the relations of students and teachers as well as their relations outside the school to other spaces constitute the living school space. The school is not a space isolated from other societal spaces but a knot in a spatial web.

Second, space is plurality; it entails the possibility of the existence of multiplicity. People are making space as a social product. There is no space before social interactions and coexisting heterogeneity with different voices. Additionally, space is constituted within different relations and connections to other spaces. From my perspective, this also includes cyberspaces and other public and private media spaces, which are important parts of the youth identity.

Third, space is never finished, fixed, and clear-cut. As a product of interrelations, it is always under construction and without clear boundaries. The school space merges with other spaces in the same way that school literacies merge with literacies of other domains.

Thus, space is open, under construction, and includes different kinds of social relations and material things, from buildings to technology and natural environments. For the school space, these propositions mean that it is lived by students, teachers, and administrators. Different people have different interests in school space and use it differently. It is always connected to home, streets, parks, and other spatial and virtual environments. We also have to keep in mind that different virtual spaces and networks are part of the school's space when students and teachers use media devices and are online.

We are always in some place. A "place" is a lived space that has meaning for us and is always interpreted again and again in processes where we live the space. Massey (1994) considered places as never-ending processes and mixtures of wider and more local social relations. As human beings, we are place-oriented and place-saturated. Therefore, places like home, school, and work have a central part in our life. But we do not so much stay in places; rather, we are on the move between places; we are mobile beings (Casey, 2009, p. xii). In fact, we can say that place appears in the move between places, "in the movement of departure and return, and in constantly crossing the border" (Rajanti, 2008, p. 78). From this perspective, a school space also constitutes border crossings and the relations to other spaces and places. Therefore, moving between places, for example from home to school and vice versa, is part of the spatial experience of home and school.

Strategic and Tactic School Space

But is a school space really a plurality and open to different practices? Is it rather a relic from the "industrial era," like Robert Pearlman and other researchers suggested? Pearlman (2009) wrote that school has changed little or not at all in 100 years:

> Hasn't it been long enough? Over 100 years of public mass education, nearly 10 years into the new century, you still see the 30-student same-look classrooms with students sitting in rows and columns listening to teachers and doing monotonous worksheets. (p. 14)

Traditional school space is constituted by the "grammar" of schooling. Grammar of schooling includes the regular structures and rules that organize the schooling, like standardized organizational practices in dividing time and space, classifying students separated by age, and splintering knowledge into different subjects (Tyack & Tobin, 1994; Selwyn, 2011, p. 88).

The grammar of schooling constitutes a structure and organization of place and time at the school. From this perspective, it is possible to see the school as a "total institution" (Goffman, 1961). By total institution, sociologist Erving Goffman means a place where people are cut off from the wider community and concentrated to the formally administered round of life. In total institutions, people are situated in the same place under the same single authority and share the same activities, like schooling, which is tightly scheduled and spatially organized. Total institutions are enclosed symbolically by walls and locked doors and characterized by a number of social and cultural structures and hierarchies. These are also present at school.

Michel de Certeau (1984) called *strategy* the logic of closure and internal administration of institutions (Lankshear & Knobel, 2011). In his theory, de Certeau distinguished between two social forces: production and consumption. Production is controlled through those who have power, who create, maintain, and impose disciplined spaces, make timetables and procedures, and organize the life in spaces. They control through strategies, which are processes directed toward disciplining places and maintaining power (Gomez, Stone, & Hobbel, 2004). For de Certeau, *institutions*—and if we use Erving Goffman's term—*total institutions*, are strategic when they distinguish their own place from an environment and discipline and dominate the space. But this does not mean any absolute domination and spatial isolation. For example, places like prisons can be totally isolated by architectural design, but the most important thing for strategies is not how architecture shapes the modes of living but how occupation of the place itself constructs the space (Buchanan, 1993).

In schools, the acts of schooling, teaching, and learning are strategies that construct the space and teachers are producers who wield the power and mediate the surveillance. Consumers, like students at school, consume the products of producers; they are constrained to operate within disciplined spaces. "Products" are not only commodities but also discursive orders that define and constrain forms of social practice. Social practices and formal learning at school are largely controlled through strategies. On the other hand, teachers are also under surveillance by school and city administrations that are seldom directly present in the everyday life of the school but set the structures and conditions of the schoolwork. For example, educational policy, the school curriculum, textbooks, and schools' technological infrastructure have strategic power over teachers.

Although, for example, the school space is controlled, it is not absolutely dominated. As Ian Buchanan (1993, para. 21) wrote, controlled space is "reactive rather than active. It is subject to appropriation: its disciplined/dominated spaces... can always be made smooth by their occupants by the act of occupancy itself." This "occupancy" is *tactic.* De Certeau (1984) spoke about everyday resistance, in which people undermine imposed power relations. He gave an example of everyday tactics referring to the French practice of *la perruque.* It is a worker's own work disguised as work for the employer. A worker, for example, uses "company time" for his own purposes and takes some time from the company "for work that is free, creative, and precisely not directed toward profit" (p. 25). In de Certeau's theory, tactics is an art of the weak. He illustrated this by explaining the practices of North African migrants living in Paris. According to de Certeau, these migrants insinuate into the system imposed on them "the ways of 'dwelling' (in house or a language) peculiar to [their] native Kabylia" (p. 30; cf. Lankshear & Knobel, 2011, p. 243). They create their own plural space to be by the "art of being in between."

In fact, tactics is not only resistance but also making things smoother and more comfortable. Tactics has these two sides but not necessarily at the same time or in every case. On the other hand, tactics helps to obtain some satisfactions and pleasures in everyday life and routines. Lankshear and Knobel (2002, para. 30) called this "the sustaining side of uses and tactics." The other side is more active and in some cases more activist. It is about resisting and tackling. Both sides of tactics have an active role in the constitution of space. They are not only reactive against strategy; they "define the limits of strategy, and inform its modes of operation in a fundamental sense" (Buchanan, 1993, para. 22).

Strategies and tactics are intertwined together: tactics open dominated places and reshape strategies, and strategies are reactions to tactics. In this way, for example, a school space is full of orders and prohibitions but also negotiations and reorientations in the space. There are always "holes" in the strategic and institutional space into which students can escape without leaving the space and make it more comfortable (Buchanan, 1993). Students make their own room in the school by different activities.

It is important to note that by tactics, students are no longer just consumers in the school space but also producers of new spaces that they occupy. One example of this is using mobile phones at school. The use of mobile phones is generally not allowed for personal purposes in class. In Finnish primary schools, students need special permission from their parents and principal to carry a mobile phone in school. Usually, it has to be turned off during the school day. At secondary schools, mobile phones are allowed but not during class.

There are many reasons for banning mobile phones, but from the perspective of strategies, banning helps teachers dominate the space. Mobile phones are like wormholes (Rajanti, 2008) to another space outside the school. Rajanti (2008) used the concept of wormhole to describe "postfordistic society" as a possibility to jump from one reality to another, from space to space, especially by using digital networks. Wormhole is a well-known concept from theoretical physics and sci-fi fantasy, a hypothetical possibility to leap from one dimension to another. In the context of "postfordistic society" that has dissolved traditional borders of private and public spaces, wormholes make it possible to move between spaces and inhabit new ones. Rajanti pointed out that by using ICT

> even in the intimacy of one's own home one is connected to other people, and not just a few chosen ones, but anybody. From the dichotomy of public and private space we move to the space of multitude. The global postfordist society is characterized by the immediately social space of multitude. (p. 80)

The blurring border between public and private also changes the school space and constitutes new social orders. Because mobile phones are banned and not compatible with schooling, students must make their own tactics for the use of these devices. Mobile phones are a technology that breaks traditional social order at school and gives students the possibility to create their own time-space and a sphere of autonomous practices at schools (cf. Takahashi, 2011).

Altogether, many media devices, like television and radio, provide an audience with the "'magical possibility 'of 'being in two places at once,'" a pluralisation of place (Moores, 2012, p. 14). But the pluralisation or doubling of place in the learning and teaching context is, of course, problematic for teachers and the traditional schooling order. For teachers, it is almost impossible to keep an eye on students' use of a mobile phone. Checking and sending text messages under a school desk takes only a few seconds, and if teachers say something, the mobile phone is already in a pocket or bag. Some teachers of the Maple School tried to "clean the space" and make absolute prohibitions, but usually these did not help. They tried to control the use of different kinds of devices in different ways. For example, computer labs were closed and there were free computers only in the school library, but these were for the use of upper secondary school students only. Sometimes the doors of computer classes were open and students passing by said that they had a strong temptation to go in and check Facebook's News Feed. Additionally, some of the classrooms had a computer but the doors were always closed during breaks. In two visual arts classrooms, there were also some laptops for video editing and animation productions.

Some teachers tried to use humor when they wanted to ban, for example, mobile phones from the classroom. In a staff room, I heard a discussion between two teachers who were concerned about the students' use of mobile phones in a classroom. Another teacher explained her method: "I always say to the students that if such an important person as I does not need one [mobile phone] in the classroom, then neither do you. Well, I try to meddle by making jokes" (from my notebook.)

Using media and different literacy practices during lessons is also a question of tactics and negotiations between students and teachers. When I was observing the visual arts teaching that was integrated into the ICT teaching, the students had an assignment to take a self-portrait by a still camera and manipulate and modify their portrait using the computer. They started photo manipulation with the guidance of the visual arts teacher. While using the photo manipulation software, students were permitted to listen to music "with another ear" with their mobile phones or MP3 players without disturbing other students. After a while, the visual arts teacher went back to his own classroom and the ICT teacher came to the computer lab. One of the boys, Janne, walked to the front of the classroom in order to collect headphones for himself. The ICT teacher asked why Janne needed headphones. "I listen to the music from the computer," Janne said. "In my class, it is not allowed to listen to music," answered the teacher. Students complained strongly because the visual arts teacher had given permission just a few minutes ago. The ICT teacher had to admit that if another teacher had given permission she could not cancel it, but she emphasized that when she is teaching, listening to music is not allowed.

This is an example of a negotiation situation. The school has rules but these are not absolute. Teachers have autonomy in their classroom and possess different kinds of attitudes, experiences, and working methods. Because mobile phones are part of students' everyday life and communication procedures with their family and other people, they bring these devices to school and everywhere they go. We can say that a mobile phone and other media devices are a part of the body and identity in the same way as clothes. This is the picture, I think, that is common in almost every school in technologically rich countries: Riku, age 15, used big red headphones and carried them on his neck almost all the time at school. In an interview, he said that he is not able to get through school without music and his MP3 player: "I get a good feeling and so.... Without music the school day would be boring. Music is important for me." Music and big red headphones were an important part of his image and identity at school. He also showed this by singing along to the music he was listening to, even in class.

Singing in the classroom and listening to music was a tactical act that also had some features of resistance. When Riku was bored in the classroom, he picked up his red headphones and started to listen to music. Teachers sometimes had to ask him to stop listening many times.

Strategic logic constructs what I call an *official school space* (cf. Gordon, Holland, & Lahelma, 2000). Official school space includes formal learning activities, classroom settings, and experiences during actual teaching and learning. An official school space has a particular organization of time, space, and interaction among people in a learning setting. David Barton (2007) described school as a particular domain that has physically distinct architecture with clearly demarcated boundaries. Within school, space is divided up as well as school time. School space and school life are marked by shared rhythms that everybody follows in certain sequences. Students gather together with one teacher in a classroom and then move from classroom to classroom along corridors where they can sometimes spend their time between lessons. In official school space, activities are broken up into subjects, "which are pursued at separate times, often in different places, with different teachers and with different books" (p. 177).

But schools also have other kinds of spaces, which I call *unofficial school spaces*. They include physical school space outside of classrooms and learning settings, peer-to-peer relations, social practices, and in most cases, popular media representations, media consumption, and production.

School Space as a Heterotopia

When the school space is an arena of tactics, it can also be seen as a *heterotopia*. Heterotopia is a concept used by Michel Foucault in his short radio lecture, "Of Other Spaces" (1967). By the term heterotopia (literally other places), Foucault referred to places where normal social orders are temporarily suspended and then reconstructed to constitute new spaces. Foucault's notions about heterotopia were not consistent or coherent but rather a new approach to space that he never articulated in a complete way.

Foucault (1967) suggested that in the history of space, we now live at the age of sites: "Our epoch is one in which space takes for us the form of relations among sites" (para. 6). Sites are connected together as networks and grids like the traffic system or the Internet, or in terms of "demography," which for Foucault meant "the human site or living space." Heterotopias are sites that are related together and reflect and contest other sites with conflicting forces. They are sites of others, like a prison or asylum, and beyond everyday life. He gave us six principles of heterotopias:

First one is that every culture has heterotopias but as varied forms and not as absolutely universal modes. But at least two categories of heterotopias are common in every culture: crisis and deviations of heterotopias. The crisis heterotopias are privileged, sacred or forbidden places, reserved for individuals in state of crisis in relation to society. Foucault's examples of the crisis heterotopias in modern society are military service, boarding school and honeymoon trip in hotels. Foucault wrote that these are the places for first manifestations of sexual virility that are supposed to take place "elsewhere," rather than at home (Foucault, 1967).

In modern society, crises are more or less replaced by heterotopia of deviation: "those in which individuals whose behavior is deviant in relation to the required mean or norm are placed" (Foucault, 1967, para. 17). His examples are rest homes, psychiatric hospitals, and prisons.

> The second principle of heterotopias is that a society, as its history unfolds, can make an existing heterotopia function in a very different fashion; for each heterotopia has a precise and determined function within a society and the same heterotopia can, according to the synchrony of the culture in which it occurs, have one function or another. (Foucault, 1967, para. 18)

Here Foucault's example is a cemetery that is connected to all sites of the city, village, and society. Almost all people have relatives and friends in a cemetery and cemeteries have always existed in western societies. But at the same time, cemeteries have changed a lot, from 18th-century cemeteries as "city centrums" with tombstones to late-20th-century suburban cemeteries. The role of the cemetery changes with changes in the meaning of mortality.

Third principle is that the heterotopia is capable of juxtaposing several, even possibly incompatible, or to one another foreign, sites in one real place. These are places where different sites converge and become intertwined together, like the stage of the theatre, the cinema screen, and the oriental garden (Foucault, 1967, para. 21).

According to fourth principle heterotopias are linked to time; to temporal accumulation, discontinuities, intersections, and phasing of time that Foucault called *heterochronies*. These heterotopias presuppose an "absolute break" with traditional experience of time (Foucault, 1967, para. 22). Examples for this kind of heterotopias are museums and libraries, which accumulate infinite time:

> The will to enclose in one place all times, all epochs, all forms, all tastes, the idea of constituting a place of all times that is itself outside of time and inaccessible to its ravages, the project of organizing in this way a sort of perpetual and indefinite accumulation of time in an immobile place, this whole idea belongs to our modernity. The museum and the library are heterotopias that are proper to western culture of the nineteenth century. (Foucault, 1967, para. 23)

Fifth principle is that heterotopias presuppose a system of opening and closing that makes them both isolated and penetrable at the same time and divides them from more freely accessible public space. Entry and exit may have different modes, like material barriers in prisons or rituals in the Scandinavian sauna (Foucault, 1967, para. 25). These operations of opening and closing are the workings of power, which operate through the control of space and time (Soja, 1996, p. 161).

Lastly, heterotopias have a function in relation to all other spaces that remain outside of them. This is described as the unfolding of two poles: either the role of heterotopias is to create a space of illusion "that exposes every real space, all the sites inside of which human life is portioned, as still more illusory," or else on the contrary, "to create a space that is other, another real space, as perfect, as meticulous, as well arranged as ours is messy, ill constructed, and jumbled" (Foucault, 1967, para. 27).

Examples by Foucault do not provide adequate support for thinking of a school space as a heterotopia. The school space is not literally the space of otherness or a closed space like a prison. But there are some other ways to think of heterotopia as well. One interesting perspective is given by Kevin Hetherington (1997), who defined heterotopia as *spaces of alternate ordering*. He emphasized social order more than marginality and pure resistance. At this kind of site, heterotopia brings together heterogeneous and unusual things without letting them build a coherent unity. For Foucault, a written text can also be this kind of site, as he wrote in *The Order of Things* (1992):

> *Heterotopias* are disturbing, probably because they secretly undermine language, because they make impossible to name this *and* that, because they shatter or tangle common names, because they destroy "syntax" in advance, and not only the syntax with which we construct sentences but also that less apparent syntax which causes words and things (next to and also opposite one another) to "hold together." (p. xviii)

Heterotopias rupture the order of things through a different way of ordering (Hetherington, 1997). Hetherington (1997) pointed out that rather than an actual place, heterotopia is "an idea, or perhaps a practice, that challenges the functional ordering of space while refusing to become part of that order" (pp. 46–47). In this way, heterotopia offers an alternative way to do things. It is not a question of marginal places but relations to other sites:

> Heterotopia exist when the relationship between sites is described by a difference of representation defined by their modes of social ordering. For example, holding a festival next to a prison would constitute a heterotopic relationship, each space being used to order the social in very different ways. (Hetherington, 1997, p. 8)

From this perspective, we can see a school space as a heterotopia: different youth media practices break the social order of traditional schooling and build an alternative order and heterotopia. This is seen especially in creative media practices. Heterotopia is a space for media education where students bring media practices to school from outside of the school spaces. Teachers take some relations to this new social ordering where students show new kinds of agency in learning settings, both formal and informal.

Because of the notion of language, heterotopias are also interesting from the perspective of literacies and media literacies. Hetherington (1997) wrote that for Foucault, writing is heterotopic "in that it offers the possibility of similitude rather than resemblance as a form of representation" (p. 50). The difference between *similitude* and *resemblance* is a key theme in Foucault's (2008) essay, *This Is Not a Pipe,* which comments on the paintings of René Magritte. For Foucault, Magritte is a cartographer of heterotopia in a visual manner. As James Harkness (2008, p. 7) wrote in "Translator's Introduction" to *This Is Not a Pipe,* Magritte's famous images do not show resemblance to any references outside the images in the same way as in the linguistics of Ferdinand de Saussure, where any word or sign does not refer to things. As is well-known, for de Saussure, words have meanings in the language system, not in any reference system. The word "dog" does not refer to any real animal; instead, it has a conceptual signification insofar as it differs from any other word, like *fog* or *dot.* Signs are arbitrary, according to de Saussure. This is an act of similitude that "multiplies different affirmations, which dance together, tilting and tumbling over another" (Foucault, 2008, p. 46). It is different than resemblance, which is supposed to have an original model, which orders and hierarchizes "secondhand" copies of it. Hetherington (1997) interpreted heterotopia as a process of similitude:

> This process of similitude is revelatory, like a collage, it brings forward the out-of-place and offers it up as a basis for alternative perspectives and orderings, revealing what is hidden among the ruins: little fragments of past, forgotten lives, found objects, strange, unsettling novel things that have a poetic wonder about them. (p. 50)

For Hetherington (1997), the process of modernity creates heterotopias as places that are ambivalent and uncertain, "either because they are new and as-yet unknown or because they are impossible archaic representations of former modes of social order that have become obsolete" (p. 68). Heterotopias are modern places where some fears have been ordered into other places, like death to the cemeteries and criminality to the prisons. Children in schools are not necessarily fearful (although there is always some peculiar otherness in teenagers, and societies have always had moral panics toward the media practices of youth), but in the same way as otherness in general, children are defined against normality. David Buckingham (2000) wrote that school and schooling as a social institution usually define what it means to be a child and a child of a particular age. Schooling not only organizes the time and space of a school but also the position of a student of a school as a child:

> The separation of children by biological age rather than "ability," the highly regulated nature of teacher/student relationships, the organization of the curriculum and the daily timetable, the practice of grading—all in various ways serve to reinforce and naturalize particular assumptions about what children are and should be. And yet these definitions are, for most part, only made explicit in specialized forms of institutional and professional discourse from which children themselves are largely excluded. (Buckingham, 2000, p. 7)

As objects of different "strategies" in de Certeau's sense, "children" are defined in official school spaces where social order is imposed. Instead, unofficial school spaces offer alternative orders where children are able to define their identity themselves by doing things in an alternative way. In this way, the school is ambivalent, a space of acts of strategies and tactics and different literacy practices where teachers and the schooling define childhood and studenthood, but the young do it by themselves as well. Mark Twain tried to describe adolescents' search for identity in the famous Huckleberry Finn story through the allegory of a Mississippi river trip, a time for Huck Finn to escape from an adult world and social order. Interestingly, for Foucault, a boat was "the heterotopia par excellence" (Foucault, 1967, para. 29). A boat is a floating piece of space going from place to place as some kind of liminal space. These kinds of liminal places are special places to explore identity outside defined life routines and the gaze of adulthood. For young people, unofficial school spaces and out-of-school literacies offer "liminal experience," an experience of being somewhere between home and school and between childhood and adulthood. Therefore, as heterotopias and alternative ways of social ordering, school spaces are also spaces for creativity and experiments, possibilities to do things differently and construct new social interactions in special, spatializing processes.

Heterotopia also indicates the crossover of formal and informal learning that has usually been seen as contrary to each other, as binary concepts. Learning at school in classroom and curricular settings has been seen as formal in nature, whereas learning out of school and other institutional educational spaces has been viewed as informal. Heterotopias as alternative social orderings are also alternative pedagogical orderings where formal and informal learning collide. A school is a space for different kinds of learning because it has dimensions outside of institutional settings, especially if students' own learning practices that come outside of schools are accepted and valued.

Official and Unofficial Spaces at Maple School

At Maple School, the typical official place-time was a classroom in 45-minute sequences. A typical classroom consisted of four or five rows of desks aligned side by side in six or seven columns. Usually, a teacher was at the front of the class telling something about the day's subject, writing on the blackboard, or showing some transparencies on an overhead projector. This is often called frontal teaching. As Kaarlo Laine (2000, p. 34) has argued, teachers have authority over the institutional social space in a school. The teacher determines a space and the students' role is to react and answer to the teacher's act and presence. Some teachers at Maple School used a computer and data projector as well, but these were only available in some classrooms.

Almost all teachers at Maple School had their own classroom, which was closed and could only be opened with the teacher's key. Doors were usually closed during lessons, but one of the visual arts teachers sometimes left the door open. The classroom is the place of teaching and learning and students rarely visit other places within or outside the school during lessons. However, media education opened other spaces during school time, for example, when students went outside the school to shoot a video or used some other spaces inside the school.

Schools are traditionally arranged hierarchically with someone in charge and statutes provided for different teachers; this is seen in space as well: for example, a principal has his own office and other teachers establish places in the staff room. Teachers are not equal when they choose their classroom or their place in the staff room. Young teachers, especially, sometimes do not even have their own classroom; they have to use what is available and move from classroom to classroom during a school day.

According to Steven Hodas (1993), traditional schooling technology, like overhead projectors and chalkboards, enhance the teacher's authoritative position as an information source and reduce the physical labor required of the teacher when communicating written texts. Moreover, this kind of learning technology constitutes the school's official space-time structure, gathering students around common texture and temporally organizing lessons and activities in the classroom (Leander, 2007, p. 27).

At Maple School, as well as at other schools in Finland, schoolbooks are still crucial learning materials. Usually, teachers give students assignments for the lesson either from the book or from other learning materials. Almost all interactions, readings, interpretations, and writings happen in this physical classroom context (cf. Palmu, 2003, p. 63). Single physical space for schooling is dominant, but some curricular media practices and assignments happen outside of the classroom as well.

Going outside of the classroom and even outside of the schoolyard during a school day is a special occasion for students, as mentioned before. I observed some cases where students used the school's gym as a place for filming a video in a learning context. For teachers, having students work out of surveillance, outside of the classroom, was difficult at times. The visual arts teacher told me that in the previous year, some students were running up and down the stairs in the school corridor when they were supposed to be shooting their video: "I said to our principal that this course should be a physical education course because students are running in the corridors."

In November 2009, 4 eighth-grade boys (Janne, Riku, Pertti, & Taisto) were editing their short video entitled *iScream* at school. The assignment was part of the formal learning setting in a movie course. The boys were alone in a small editing room next to the visual arts class, with three laptops connected to a Wi-Fi network.

> Riku has his headphones on and is singing a song by the Black Eyed Peas: "*I got a feeling that tonight's gonna be a good night....* "
>
> Janne is trying to find sound effects for their video on the Internet, using Google. "Hey, shut up!" he shouts to Riku.
>
> Riku turns from his laptop: "I'm on *IRC-Gallery* but I'll close it. There's nobody. Do you want audio files? I can help you." Riku puts his headphones around his neck and turns up the music so he can hear it through the headphones without putting them over his ears. The boys try to listen to the sound effects from the *Soundsnap* website to find effects for the gunshot in their video but don't find the right one.
>
> "This sounds like somebody bouncing a basketball," says Janne.
>
> Riku turns back to his computer and goes to his profile on *Aimini.net*, a site for sharing music, photos, videos, and games. Pertti is on *Facebook* and Taisto is trying to help Janne to find the right sound effects. Riku finds some music on the laptop and tries to download it to his mobile phone. "I don't have a cable. I can't download. What kind of cable do you have?" he asks Janne.
>
> "Sony Ericsson," Janne answers.
>
> "Phew, some Sony Ericsson," says Riku. He opens the game forum *Arcade.fi* and starts watching a Flash movie: "Really good work. It would be really nice if this were a game."
>
> "You can't make a Flash game work like that. It would need so much scripting it wouldn't be a game anymore," says Janne. At last Janne finds the right sound effect for their video: "Here it is, would you please turn that music off!"

he tells Riku again. "That's great." He connects his mobile phone to the laptop via a USB cable and downloads the sound effect to his phone.

The boys have to change laptops because their movie is on Riku's computer. Janne opens *Final Cut* for editing and starts to transfer sound effects and edit the movie. Riku and Pertti start to play the game *Free Rider* on *Arcade.fi*. (From my notebook)

The situation just described is completely different than normal, when the teacher is teaching in the front of the classroom and students sit in rows and blocks. Four boys are working together and the teacher is missing. The situation is more like hanging out with peers, playing and having fun, but still they are doing their schoolwork and are in the school in a formal learning setting. Making media and different media practices are examples of tactics, how students find their own place in the school and make the space smoother. The space outside of the official school space is students' own space where they are free to use their own media practices.

This is an example of unofficial school space and heterotopia of the school, an expression of other spaces and social orderings like some in-between spaces between lessons. I observed young people spending their time in corridors and the schoolyard or outside of the school area during their breaks. They were in small groups and just hung out, listened to music, sent text messages, watched movie clips, and took photos. Teachers did not rule this space structure in the same way as official school space.

Official and unofficial school spaces have some especially interesting differences also in terms of time. In the official school space, time is understood to be *monochromic*, meaning that people use their time on a particular assignment until the assignment is completed (Jones, 2005). In an unofficial setting, particularly when online, young people mix and blend timescales and activities. Their temporal orientation is different. Young people multitask, doing different assignments at the same time.

In the aforementioned situation, the students are multitasking. Riku in particular is switching from one task to another. His orientation to time is *polychromic*. We can also say that their experience of space here is *polyspatial*. Riku and the other boys have different possibilities for orientating their attention in terms of space. Jones (2005) wrote that in computer-mediated communication, there are at least five spaces: (a) the physical space in which people are acting, (b) virtual spaces created by the interfaces they are using, (c) relational space created by the state of talk among participants, (d) screen space (the actual space of the users' screens), and (e) third spaces (spaces referred to in the course of interaction). While the boys were doing their school assignment, they shared some of these spaces while also having different screen spaces and virtual spaces.

It is significant to recognize that the media practices of the youth described here constitute a different space-time structure from what is common in formal learning practices in an official school space. Kevin Leander (2007, p. 27) has argued that the problem of integrating the Internet into schooling is not technical but rather spatial and temporal. Leander (p. 47) wrote that the space-time typically produced in schools involves sequential activity where everyone follows the same sequential path, a single space for each assignment and authorized texts and space. By contrast, in online space and online practices, simultaneous activity is the norm, practices happen across multiple spaces, and multimodal texts and spatial authority are not strongly held. Online practices do not fit within traditional schooling because of this different orientation in time and space. Nevertheless, young people bring practices of this kind into their school space-time and organize it in a new way.

Media in and outside the School

Traditional schooling inside classrooms was normative in the Maple School but, of course, not the only way to organize teaching and learning. Media education especially, as well as working and practicing with the media, transformed traditional time-space structure. My ethnographic fieldwork started from a special place outside of the school. Ninth-grade students visited an art museum where a

Figure 3.1: Everyday Media Usage ($n = 305$)

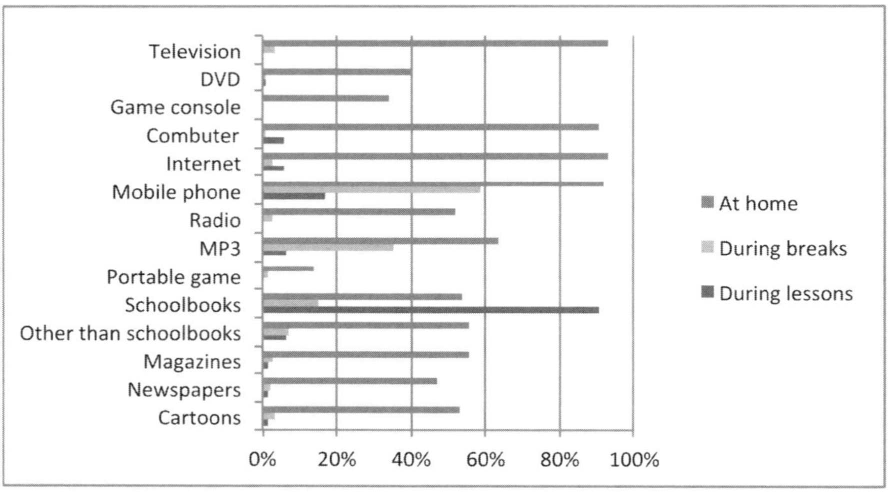

Finnish media artist had an exhibition. Visiting an art museum was part of the visual arts class and the teacher had visited the exhibition before with the eighth-grade class. It was a hot summer day. When I stepped into the museum, a class from another school was there too. The Maple class arrived at the museum a little later. The class started from the back of the museum where the teacher introduced the exhibition to the students and asked questions about media and media art. One of the students defined mass media as an example of media. "Media art" was not so simple, but students found video and animation as part of the scene. The teacher used an article from Wikipedia to define media art, which includes, among other things, videotape, multimedia, network art, light and sound art, and an installation.

Media as such is an essential part of young people's everyday lives, but the museum is not so familiar to the students. Inside the museum, one of the boys commented that, "In the museum I get sick; it happens in all old places." For him, a museum automatically means something ancient. On the contrary, media and especially digital media are something new and interesting. It is possible to see the same dichotomy between the school and students' everyday lives or between school literacies and out-of-school literacies.

School literacies are mostly intertwined with books as an essential part of schooling. My survey showed that besides books, any other media device, such as a computer, television, or magazine, is hardly used during school lessons. Ninety percent of students said that they use schoolbooks daily during lessons at school. The next most common media device was surprisingly a mobile phone, but it was not used in a formal learning context (see Figure 3.1).

Figure 3.1 gives a general picture of the media ecology and devices used daily by students at Maple School. It shows a large difference between school and out-of-school media. Mobile phones, the Internet, computers, and television play a great part in the lives of the youth but mostly at home or outside of school. The interviews confirm the high significance of these devices. But there were also strong differences between the youth. Some of them read 70 fiction books in a year, while others read none. "I do not read books because they are boring" wrote one 14-year-old girl in her media diary. Some sent over 1,000 text messages in a month and others used the mobile phone mainly to make calls. Some were heavy video game players; others played only occasionally. Most of the teenagers had a profile in some social networking site, but some did not have any at all. Some blogged or wrote fan fiction and published photos on a social networking or some other website.

Anna Burke and Roberta F. Hammet (2009, p. 3) argued that young people's new media environment and practices have given rise to tension or bifurcation in school and out-of-school literacies, where school literacy is understood in narrow forms of written text. School literacy is said to be totally different than kids' everyday out-of-school literacy, which deals with a multimodal composition of different combinations of sound, text, images, video, animations, and graphics. Lankshear and Knobel (2007) argued that old modes of schooling include individual productivity, competence, and intelligence, whereas new digital environments prefer participatory, collaborative, and distributed competences and literacies. They spoke of the "deep grammar" of schooling, which constructs learning as teacher-directed and curricular and cuts schools off from new literacy practices (Lankshear & Knobel, 2006, p. 55).

From this point of view, school literacies and out-of-school literacies are far away from each other. On one side of the abyss is the teacher-oriented "industrial view of knowledge building," and on the other side is "post-industrial collective intelligence" (Lankshear & Knobel, 2006, p. 55). Sometimes it is thought that there has to be some bridge between these poles. Kate Pahl and Jennifer Rowsell (2005, p. 66) argued that this is the "third space," where students are allowed to use both school and out-of-school literacies.

The term *third space* was originally used in hybridity theory (Bhabha, 1994). Hybridity theory "posits that people in any given community draw on multiple resources or funds to make sense of the world" (Moje et al., 2004, p. 42). Moreover, for Bhabha, cultural hybridity as a third space means the in-between space, which brings together these contradictory knowledges, practices, and discourses.

Even though schools are in many cases ultraconservative and old-fashioned in the eyes of young people, students bring their own media practices and literacy practices to school. In this way, a school space is not one-dimensional but rather fosters cultural hybridity. A school is not a closed, undialectical, monospatial, and immobile space but a multiple, dynamic, open, and hybrid space-time that has different relations to the students' everyday practices and learning.

Students can use their everyday practices as tactics in order to transform school spaces. In the same way, as we have seen, all teaching contexts are not necessarily the official school space. Students and teachers are moving between the places. New media technology even helps to do this. For example, students are able to use mobile phones in order to locate themselves and find alternative spaces for learning.

Digital Technology in Official School Space

After all, implementing ICT and digital technology in learning is still in the beginning. Finland is a technologically rich country where schools have relatively good technological resources, but the use of information and digital technology in school settings is lower as compared to most active countries (Kankaanranta & Puhakka, 2008). In the 1990s, Finland was known as one of the leading information societies in the world. The government strongly supported the educational use of ICT, but advancing ICT use in schools was mainly based on different projects and was supported by extra funding (Niemi, Kynäslahti, & Vahtivuori-Hänninen, 2012). Hannele Niemi and her contributors (2012, p. 2) wrote that numerous projects did not have "a broad or lasting impact on the educational performance of schools."

The much promised technological "revolution" in learning and teaching has failed to materialize (cf. Selwyn, 2011). Strong investments in technical resources and infrastructure do not automatically create new teaching and learning cultures (Niemi et al., 2012, p. 2). Neil Selwyn (2011, p. 25) referred to numerous surveys and reports, which confirm that even though the physical presence of digital technology and media devices has increased at schools in recent years, the use of technology in school practices is limited, far away from the "real-world" technology use and often focused on matters of school management and not so much on teaching and learning. Bigum and Rowan (2008, p. 249) argued that the use of technology at school is often a "pretend," "fabricated," and "inauthentic" version of the digital practices that take place outside of school. Digital technology use at school is largely formalized and bounded by nature (Selwyn, 2011, p. 25).

At Maple School, almost every classroom includes a computer, which is mainly for the teacher's use. The school has two computer labs. During my fieldwork, the other computer lab got new computers, but according to the vice principal, it was difficult to get the new ones:

> For two years we have got promises but then came this financial breakdown and promises were canceled. We got stuck in the hole. There was supposed to be all systems in all classrooms but then we got nothing. (Interview with the vice principal, female)

After a long struggle, Maple School finally obtained new computers for the computer lab, along with new software. As the vice principal said, "We have to think with the students how everything works with new software and applications." She did not have time to study the new software, but fortunately there were some students who knew something about it and therefore they studied software options together. Thus, in Maple School, students sometimes help teachers solve problems, and in official school space, their roles have turned upside down: students teach teachers. The principal gave me an example of a seventh-grade student who helped him use ICT in a classroom:

> We had to go to the upper secondary school classroom and there was this equipment and I didn't understand how to use it. I asked from the class if there is any expert. One boy said that he could come and see if he could help me. It didn't take a long time and he gave me advice and everything worked well. (Interview with the principal, male)

The principal said that he referred to the same boy when he had problems with ICT in other classrooms as well.

In a visual arts class, students were using old laptops and an old wireless network. Some years ago somebody had called from the local school administration and asked if there was any need for Apple iBook laptops at the school. Nobody inside the school had planned this kind of digital infrastructure, and as one of the teachers said, laptops just "came in." From the beginning, the data storage of the laptops was too small and the laptops were not suitable for the school's practices, like video editing and animation. Additionally, teachers were used to using Windows computers and not Macs.

The problem was that teachers were not involved with the planning of digital technology use in the school. They, more or less, had to conform to a decision that was made outside the school without sufficient pedagogical and local expertise. Decisions were made by school administrators somewhere in the bureaucracy.

Teachers and students are not free to choose any kind of teaching and learning material they want to use. In the Finnish school system, teachers are independent when they plan their teaching, but at the same time, schools are subject to large-scale administrative organization where services are planned by a central administration. Like schools themselves, schools' technology is "a knot that is 'tangled up' in a web of practices that stretch into complex systems beginning and ending outside of the school" (Selwyn, 2011, p. 119). Technological problems or the use of technology in an official school space is tied to the complex structure of schooling:

> The paucity of digital technology use in schools is not due simply to logistical deficiencies such as a lack of resourcing, whole-school "vision," technical support or time. Neither are these "problems" caused by deficiencies on the part of individuals working within schools. It is not the case that teachers and school leaders simply "do not get" digital technologies or lack the necessary skills, aptitudes or ambitions to make the best of technology in their work. Rather...digital technology in schools is the preserve of a wide-ranging complex of interests—from state policy-makers, IT firms and advocacy groups to schools, teachers, parents, and (of course) the students themselves. (Selwyn, 2011, p. 120)

Niemi and her research group (2012) found that for the successful implementation of ICT in the school, ICT should be adapted as a convenient and natural part of everyday school life. In Finnish schools where ICT was used successfully, it was not a separate part of the school environment but integrated into all schoolwork, using a joint vision of principals and teachers. "The schools shared a vision and values, and respected teacher teamwork and a culture of sharing" (p. 7). Also, students had an active role in knowledge creation and participatory learning settings. Niemi and her contributors (p. 12) concluded that many components have to support each other in order to successfully implement ICT in the school culture: it requires both pedagogical and organizational qualities, management, investments, and flexible curriculum.

But "joint visions and values" are not so easy to achieve. Operators in a school sector usually have their own ideas, ends, solutions, and evaluative means for educational technology. Usually, technological infrastructure in schools is planned by nonteachers and inserted into the school from the top down, often as part of national school technology policies (see also Selwyn, 2011, p. 33). As Selwyn (2011, p. 33) argued, "dumping" digital technology on schools does not meet the needs of local schools and teachers. The situation is frustrating for many teachers. One of Maple School's visual arts teachers described the situation there in the following way:

> I had five or six years ago clear plans for what we need for media education. For example, that we should get more devices every year so that it does not take all our money in one year, for example one or two laptops per year so that we could get eight laptops in both visual arts classrooms. But the practice is different, we got one laptop for a teacher in one year and that was all. Next year we didn't get anything and we were forgotten. Then we should do new plans and again other new plans. It makes me cynical, why we had to do new plans for nothing? (Interview with the visual arts teacher, male)

He added that for 2 years, he studied certain computer software for video and image editing but never received that software in the school. Administrations create certain strategies (in de Certeau's sense) and teachers and students have to conform or use the technology in their own tactical way.

Another teacher who worked as a media pedagog, offering support for media education in the city area, commented that the idea at the school administration level has been that people have to move to the places where technology and media already exist, such as computer labs:

> But I think that the idea should no more be that people have to move to different places to watch how technology works but technology comes to the people and sees how people work. (Interview with the media pedagog, male)

This point of view supports the idea that technology should be a natural part of everyday school life, just like it is in the lives of youth.

Technology serves different practices, and it is not necessary to establish long lists of what somebody has to do when using a computer and certain software. "In fact, a student does her own things and you cannot have a strong effect on her practices," the media pedagogue argued. After strategy comes practices and tactics, where students use media and technology in their own way. Teachers can give some tips, but strong strategies are not effective in media education.

Official school spaces and strategies offer technology as some kind of matrix that everyone at the school has to follow. Students are even socialized into formal ways of using digital technology in official school spaces. But school technology does not automatically fit to the everyday school practices and spaces. There are several issues that affect problems when technology is used in schools. One of these is time.

From the teacher's perspective, time is the scarcest resource in schools. One of the mother tongue teachers commented that 45 minutes is an extremely short period of time, especially if the class has to do something other than listen to the teacher:

> You dig up some peripheral material, share it to the students and soon after you have to gather it back. It is really frustrating sometimes. (Interview with a mother tongue teacher, female)

Lack of time was sometimes an impossible obstacle to the use of ICT at school. For example, in a visual arts class, the teacher tried to solve a problem of how to upload photos, taken by students, to the computer:

> We have in our classroom one computer and every student should upload a photo to the school's network. It takes five minutes per student. We have 24 students. You can count how much time it takes. (Interview with a visual arts teacher, male)

It is easy to see that official time-space in the school does not always allow an effective and creative use of digital technology.

At the same time, digital technology is something that sets some new demands for the official school space and teachers. When Maple School had its anniversary in 2008, a display was set up in the school hallway. It was designed for the students' project presentations, but there was nobody who would update the software and compose new presentations. Teachers did not have time to edit videos and other material in order to present students' media productions within the display. A visual arts teacher commented that there was a strong need for a media assistant whose task would be to service media devices and collect students' work for presentations. The principal had even calculated that a media assistant would not be extremely expensive, especially if he or she would serve several schools, but the school administration was not interested in hiring anyone for this type of position.

Digital Natives and Immigrants at School

Some teachers had also adopted a role of digital immigrants and had some difficulties with school technology, but students helped them solve problems. For example, I observed a case where a teacher tried to use a DVD player but could not change the language in the language settings. Students started to get restless and shouted some advice: "Click the menu using remote control."

> *Teacher*: I do not have this kind of device at home. You [students] are a different generation.
>
> *Girls*: Hey, guys, you can be gentlemen and go to help the teacher.
>
> *Teacher*: It is dark in here. (From my notebook)

At last, one of the boys went to help the teacher and the DVD screening started. Students commented about the situation later, pointing out some incapabilities of teachers. The visual arts teacher commented: "Teachers' media skills are different than yours." Teachers sometimes used this "different generation" argument and received help from students. The students were pleased that they were able to handle devices that the teachers could not. Sometimes, the teachers just did not have time to learn how to use devices and software, or they were in a classroom other than their own and were not familiar with the devices.

Some scholars have argued that school has faced a "legitimacy crisis" with the students, especially because of the digital divide between teachers and students (see Selwyn, 2011, p. 31). However, it is too much to say that the whole problem is between generations or "digital natives" and "digital immigrants."

The digital natives/digital immigrants discourse provides a slightly distorted picture of the gap between students' and teachers' digital orientation. The notion of digital natives has generated strong debate since Marc Prensky (2001) popularized it in his essay, "Digital Natives, Digital Immigrants" (cf. Thomas, 2011a). After 10 years, the term has seen disdain as well as acceptance. The most optimistic views are based on "techno-evangelism" and on three main assumptions (Thomas, 2011b; cf. Bennett, Maton, & Kervin, 2008, p. 777). According to these assumptions, young people are said to (a) constitute a homogenous generation that speaks a different "technological" language from earlier generations, the digital immigrants; (b) have different learning styles from preceding generations of students; and (c) demand a new way of teaching and learning.

David Buckingham (2011) argued that these kinds of views typically overstate the extent and effects of technological change and differences among generations, while understating the diversity within them. There may be as much variation within the digital native generation as among the generations (Bennett et al., 2008, p. 779; Levy & Michael, 2011). For example, we must keep in mind that the frequency and nature of young people's Internet use differs among age groups, socioeconomic background, and other demographic variables (Bennett et al., 2008; Livingstone, Bober, & Helsper, 2005).

My study indicates that some of the students were not at all interested in using the Internet, digital media devices, or computers. I sometimes heard students screaming "I hate computers" as well, especially when problems occurred. In one interview, a 14-year-old girl told me that she does not watch TV or have a profile on any social networking site: "I do not use computers much. I do not have time." She mainly used computers for schoolwork, read headlines and comics from a newspaper in the mornings, and sometimes listened to the radio. She used her time for other hobbies outside the home.

As much as there is variation among young people and students, there is variance among adults and teachers. During my ethnography, I interviewed teachers who really tried to use digital technology in a new way but ultimately faced obstacles. For example, for some teachers, the lack of digital technology at school was frustrating. One of the teachers was a writer of a new schoolbook that included digital teaching material for teachers, but she did not have a computer and data projector in her classroom and could not use her own published digital material. The school administration tried to get a laptop and portable data projector to the classroom. Another teacher could not use his PowerPoint slides and other digital resources because he always had to move from classroom to classroom (as the youngest teacher in the school, he did not have his own permanent classroom). Other obstacles included lack of time and space, broken technology, and some organizational and infrastructural problems.

Some teachers did not get enough technical support and did not even have permission to update computer software. Sometimes a teacher and students spent almost a whole lesson trying to solve some technical problems without any technical support. It was extremely annoying when students had worked hard for hours, sometimes through the night, to get their media production ready for a presentation only to face technical problems in the classroom just before the presentation. The presentation and "premiere" of one's own media production in front of the whole class is an important part of media education. It is a time to share the feelings of success and appreciation, but total failure due technological problems does not raise motivation for the use of school technology. Unfortunately, the students seemed to get used to facing such problems and adversities.

Unofficial Technology at School

In official school spaces, the technology is primarily hierarchical, controlled by the teachers and especially by the administration. As we have seen, talking about ICT in schools usually means computer-supported learning environments. Presently, some schools in Finland have started to consider using computer tablets in classrooms. But ICT in schools seldom refers to portable devices like mobile phones, game consoles, MP3 players, or digital cameras.

In Maple School, some teachers tried to find creative ways to utilize students' own devices in formal learning settings. For example, the visual arts teacher said that sometimes mobile phones were used when students made drawing sketches outside the school. The teacher advised students to take a picture with their mobile phone so that they could continue to draw in the classroom. On the school's "physical exercise day," when every student had to choose some form of exercise, from long distance walking to soccer, one option was "statue orienteering," where the assignment was to walk a route where the student could find public statues. Students had to have a camera with them to take photos in order to prove that they had visited every statue, which is where cameras on mobile phones came in handy.

These are only a few examples of how students and teachers find new ways to use portable devices in formal learning settings. For example, mobile phones can be helpful devices, especially for creative schoolwork, because new mobile phones include a camera and microphone, which offer possibilities for media productions and audiovisual integration in the classroom. However, these are not the only possibilities. For example, the GPS logger of a mobile phone can be used in physical education and geography. The possibilities are unlimited.

Outside of the school and in unofficial school spaces, these kinds of devices are personal and sometimes used in very unconventional ways. My example is 14-year-old Eve who publishes photographs, drawings, and fan fiction at the deviantART online community. She is a manga and anime fan and uses her mobile phone for fan fiction writing in school spaces. In an interview, she told me about her practices:

> *Eve*: In school, when I get an inspiration, I write with my mobile phone and later on the computer and from that to deviantART.
>
> *Interviewer*: Do you write with your mobile phone?
>
> *Eve*: Yes, on my memory card, there is lots of free space.
>
> *Interviewer*: Where do you do it?
>
> *Eve*: Everywhere, for example if I have nothing to do during breaks at school, I take my mobile phone and write to the notebook. I write everything with my phone keyboard and then save it. Later, when I think that I could write something for deviantART, I take my phone and copy my texts onto the computer. (Interview with Eve, 14, female)

Writing fan fiction using a mobile phone is not a common use of the device. Eve did not have the possibility of transferring her writings directly to her computer, but she copied her text and rewrote it on the computer manually.

The most common way to use out-of-school technology in school spaces was by communicating with other friends, family members, and schoolmates by text messaging, phone calls, and status updates on social networking sites. This practice opened the school space to a more public sphere where students commented about what they were doing at school and what feelings they had. In January 2010, 14-year-old girl Gomi updated to her weblog

> **Nothing**
>
> We have now movie course and we can't see our video files on the computer and therefore I am now on the Net.
>
> Our video is bad and stupid because there is no actors or anything. I even can't use Final Cut ;)
>
> So, it's going well. (Gomi's weblog in January 2010)

Gomi was frustrated and expressed her feelings by publicly uploading a blog entry to the Internet during class. She inhabited at the same time both official school space and more public cyberspace by using her literacy practices. Sherry Turkle (2011) wrote that people are moved from "multi-tasking to multi-lifing," doing some kind of "life mix" between offline and online spaces at the same time. This is possible because entering virtual space is not bounded to any special place and time. Mobile devices are portals to move into virtual spaces everywhere and everyplace. People create "online selves" that are also distinct personalities in some sense when they live their own life on the World Wide Web. For example, Gomi worked hard in the movie course, but at the same time, her "avatar" made personal comments on the public web spaces.

Students' personal digital media at school is some kind of alternative media for their personal life but far away from the official school media and formal learning. Sometimes they tried to find ways to use their devices in the formal learning context, for example seeking some information from the Internet or using their mobile phones for sound recording, but this was not common. More common was communicating with others and sharing some fun stuff and listening to music with headphones.

Consuming the Media

Different media practices were important for students at school, especially during breaks (see Figure 3.1). Janne, a 14-year-old student, told me that he watches movies from his iPod during breaks. A lunch break is a good time for watching movies; if he eats in 10 minutes, he has half an hour for a movie. Consuming media is a very common way to spend free time at school, and that includes, for example, reading books, listening to music, and watching YouTube videos via mobile devices.

Consuming media at school is not, after all, only an individual habit to spend time; it has a social value of belonging to the group and relations to the whole popular culture as an important part of identity and personal taste. Caroline Dover (2007) wrote that when her research group recorded 83 hours of spontaneous interaction in two London secondary comprehensive schools, they identified 275 episodes where students talked about popular culture products like music, television, films, computers, electronic games, newspapers, and magazines. Dover argued that "media products enter into everyday conversation, facilitating relationships and acting as symbolic makers of taste and identity. Furthermore, they are present in an environment that would seem to exclude the relevance of media texts" (Introduction, para. 13). Media and popular culture are also a part of the funds of knowledge in young people's life at school. Jan Nespor (1997, p. 174) wrote that children *exchange, invoke, inhabit, and appropriate* media culture presentations that are circulated through the spaces of popular culture. Engagement with popular culture is a fundamental social activity at school and is situated in different school spaces. Here are some examples from my ethnography:

First vignette:

Subject: Movie course, eighth grade. Students are waiting for the teacher to begin the class. The teacher is helping two boys with video editing and other groups have nothing to do. One boy has a laptop on the table and he starts to talk about a video game called *Call of Duty* with his classmate. The boys have played this game together online. The boy with the laptop opens the *Call of Duty* Wiki website (the laptop has access to a wireless network):

Janne: What's your ratio?

Riku: 250 something and up. It's not worth playing with your friends on the Internet. It sucks when you try to raise your ratio. I had already 400.

The *Call of Duty* Wiki is open and the boys are watching "Callsigns" features.

Riku: I've got that one and that, and that. Wait. Then I have "Dish the Rock" and "Sidekick," "Flyswatter," "Sureshot." Wait. I've got "Preemptive Strike," at least "Iron Lungs," "Low Profile," "Grassy Knol," "UAV Jammer"—no, no I don't have that. Then "Destroyer," "Backfire," "Shot Down."

Janne: I have the "Quickdraw."

Riku: I think that I've got nothing of these, "Blindfire," that is one I have.

Janne: Yes, me, too.

Riku: Then I have "Rain of Fire."

Janne: Yeah.

Riku: Then "Bombshell."

Janne: Yes, me too, I have "Mad Bomber." (From my notebook)

This conversation and inventory takes 10 minutes before the teacher has finished with the other boys and starts teaching.

Second vignette:

Subject: Mother tongue, eighth grade. Students are in small groups. Petri, Mira, Carola, and Lauri are trying to get some ideas. The assignment is to design an advertisement for a product entitled *Jyty.*

There are music magazines for teenagers in the classroom, entitled *Suosikki.* These magazines are learning materials. Petri starts to read *Suosikki.* He reads

> an article about Sonata Arctica, a Finnish heavy metal band. "This is a very good band," says Petri. On the front page, a headline reads: "Why is *Twilight* so popular?" Mira sees the headline and says: "Because there are such good looking boys in it." Mira and Carola start discussing *Twilight* movies. Lauri utters that he has read the *Twilight* novel and asks Mira if she has seen the *Avatar* movie. The teacher comes in and asks: "Is there some extra material in *Suosikki*?" Petri says that he is trying to find inspiration. (From my notebook)

The aforementioned Mira is a 14-year-old *Twilight* fan who has read all four *Twilight* books many times over and of course has seen the *Twilight* movies. I saw her reading books even in class, whenever she had the slightest opportunity for reading. The act of reading a *Twilight* book is a kind of sign: the reader digs *Twilight* and belongs to the fandom.

These vignettes show that discussion about media taste is important for students. It provides the opportunity to enhance identity and express one's own media consumption, habits, and emotions. Game ratios, all kinds of features, devices, and manners are important social topics of conversation. Other students knew that Janne is pretty good with computers and sometimes asked for his help.

Both of these vignettes showed that social distribution of media culture is informal most of the time. Students discussed their media consumption and tastes, but when the teacher arrived, they usually stopped. But media culture is also present in formal learning. For example, in an interview, two mother tongue teachers said that *Twilight* is very popular among students, and they make many presentations about it in class. "When we taught genres of horror novels last year, *Twilight* was present in every class. Whenever a new series starts, we notice it immediately in the subject of mother tongue."

In mother tongue class, students were allowed to make a presentation or write about their own favorite movie, book or television series, "if it fits the teaching, for example how to make a review or presentation or content," the mother language teacher said and continued:

> Students gladly state their opinions and are experts in their own area. We have a video projector in the classroom, so that students can show us examples. When we taught movie genres students showed movie trailers, examples of their favorites in particular movie genres. We saw *Twilight* many times. (Interview with the mother tongue teacher, female)

Media culture is present in school in many peer-to-peer discourses but also in discussions with teachers and in schoolwork. Some teachers encouraged students to use content from popular and media culture, especially in visual arts and mother tongue classes. But these are not the only school subjects where popular culture was present. For example, in music class, students sometimes played *Guitar Hero* on PlayStation, and in physical education class, they tried street dance. The school did not have its own game console but the console was borrowed from the teacher's home. Foreign language teachers also used movies to teach language, and when students studied economics, they had the opportunity to design a business plan by creating a website. Some students said that they heard from their friends in other schools, that students in biology class tried to write poems or rap texts, for example, about cells, mitochondrions, and ribosomes.

Media and popular culture can be a part of pedagogy and help students feel more at home when they are using their own media practices, for example, playing *Guitar Hero* in class. This is a way to make an alternative space in the official school space. These kinds of practices in the official school spaces were not done on a daily basis but were rather arbitrary.

Super Communicators at School

Media consumption is not the only way that media culture is present at school. Maybe more important for the youth is to make connections to others and share feelings during the day.

Amanda Lenhart and her research group (Lenhart, Madden, MacGill, & Smith, 2007) called teenagers *super communicators*, who use any kind of device for communicating with friends, mostly by mobile phones, instant messaging applications, and social networking channels. Surveys in the United States show that mobile phone texting has become the most important channel of basic communication among teens. Daily text messaging has increased rapidly: 38% of teens in the United States texted daily in 2008 with people in their lives, 54% in 2009 (Lenhart, Ling, Campbell, & Purcell, 2010), and 63% in 2011 (Lenhart, 2012). In 2009, one in three of teenagers were heavy users who sent more than 100 messages per day, and 43% of all teens who take their phones to school say they text in class at least once a day or more (Lenhart et al., 2010).

During my ethnography, I observed that a mobile phone is one of the most important technical devices that students use every day, also in school and during school lessons. Although the use of mobile phones was forbidden during lessons, 17% of students reported that they use their mobile phone daily in class. Fifty-eight percent reported that they use their mobile phone daily during breaks and 91% at home or during free time (see Figure 3.1). Only schoolbooks were used more than mobile phones during school lessons. But this does not mean that mobile phones are an inseparable part of formal learning. Rather, they are part of students' everyday communication, identity, and media practices.

Rich Ling (2004) and Gitte Stald (2008) have articulated that the meaning of mobile phones goes beyond its practical function: it is a technical device for communicating, but this includes social practices and identity work. Additionally, new mobile phones or smart phones are much more than communication devices. They are, for example, an important medium for social networking.

Stald (2008, p. 146) emphasized that young people discuss the meaning and importance of mobile phones, especially in the social context. They find it extremely difficult to be without their mobile phone, and that says a lot about its meaning. In my interviews, almost every student confirmed the importance of the mobile phone. Sixteen-year-old Marja said that she had lost her mobile phone on the morning of our interview and that she would rather be late for school than leave her mobile phone at home. She said that she got her first mobile phone when she was a first-grade student and her phone has been her social safeguard since then. All important communication happens via mobile phone:

> It just has to be....If I have a feeling that I have to talk with someone, I call. When I am alone in the town I call somebody. Always when I feel lonely I have at least something. (Interview with Marja, 16, female)

The mobile phone brings friends near and provides Marja with the constant ability to catch up with someone. The mobile phone is the most important social communication device for her, more important than any other media device, as she explained.

Turkle (2011) wrote that young people cultivate their collaborative self by using technology as a means to be connected to their friends: "The necessity is to have someone be there" (p. 177). Super communicators are connected and share their troubles and happiness with others. Turkle interpreted this necessity as a symptom of fear of isolation and abandonment in the modern age, where many children feel cast off because of broken families and busy parents. Mobile devices and the Internet offer other communities when families are absent. Turkle suggested that this kind of collaborative self is narcissistic in nature, not indicative of people who love themselves but a fragile personality that needs constant support. Saying that continuous connectedness is narcissistic may be extravagant. Rather it is a sign of the so-called risk society.

Risk society is a term associated with sociologist Ulrich Beck (1992). Risk refers to rapid changes in society where almost nothing is stable and secure. For Beck, risks are unintended consequences of modernity, hazards, and insecurities present in modern society. Within modern families, risk anxiety has become a constant part of everyday life. The Internet as a new social space is also usually seen not only as possibility but also as a risk, where children may face different harms. But harms may be faced offline as well: traffic, strangers, drugs, and so forth, are seen as risks that children have to learn to cope with. A mobile phone may be a "safety device" in order to cope with risks and feel safe. When I asked students why they keep their mobile phones on during school classes, some of them answered that their parents want to be able to catch up with them during the school day. Children are expected to answer when their parents call and do things that are not possible without a phone. Parents want to know—and check—that their children are safe.

Hanna, 15, said that a mobile phone is not so important for her, "but I carry it everywhere with me [laughing]." It just has to be. Young people, as well as adults, are pretty much tethered to online life and their friends. Everybody expects friends to be available and answer messages right away. Emilia, 14, mentioned that a mobile phone is the most important media device for her: it is "all the time at hand." She laughed when I asked if she also uses it during class. She said that she sends about 700 text messages in a month to her friends: "I just have always to answer to my friends. Usually I do not start to send anything, but I have to answer to a friend." Emilia is always ready to answer and is connected to her friends in the same school. Using Turkle's term, she has a collaborative self.

Instead of Turkle's psychoanalytic interpretation, another reading of collaborative self may be done from the perspective of collaborative culture, where sharing ideas, emotions, and thoughts with each other is crucial. From this point of view, the collaborative self is not just fragile but open to cultural exchange of meanings. This cultural phenomenon challenges traditional concepts of private and public. In a traditional classroom setting, teaching and learning are split up into the public and private sphere: teachers give lectures or guide discussions (public sphere) and students are isolated, privately writing down notes and listening to lectures (private sphere). "The goal is to transmit information in a public way to the private minds of students" (Thomas & Brown, 2011, The New Collective, para. 2). New media technologies are making physical boundaries more permeable and transgressing the traditional boundaries of private and public (Meyrowitz, 1985; Morley, 2000; Nikunen, 2010). Mobile phones are a good example of the technology that opens schools' traditional one-dimensional communicative space to dialogical space but mostly as a heterotopia and in an unofficial way. Communication and sharing happens away from teachers' eyes, "under the table," like Emilia said. In a way, mobile phones overlay the physical locations of the media users (cf. Nikunen, 2010, p. 75). Mobile phones allow students to keep in touch with friends, surf the Internet, update status information on social networking sites, listen to music, view photos, and so forth. As a small handheld device, it is easy and unobtrusive to use everywhere and all the time in school.

School and Social Networking

Social networking is another important practice of super communicators. Facebook is the most popular social networking site among young people. According to EU Kids Online survey, 82% of 9- to 16-year-olds in Finland had a Facebook profile in 2010, as well as other social networking sites (Livingstone, Ólafsson, & Staksrud, 2011). While social networking was strongly prohibited during formal learning, students in computer class often visited their profile and uploaded comments and status updates while the teacher was not looking. Visiting social networking sites was also possible from the mobile phones of some students. Two 15-year-old girls[1] told me that a mobile phone is important because they have access to Facebook at school and during lessons.

Interviewer: Do you update your [Facebook] status by phone?

Panda96: No, I just check what is happening—friends' updates—that's all I do.

Interviewer: Is there a lot going on in Facebook during a school day? Aren't your friends supposed to be at school?

Panda96: Yes, they are also logged into Facebook with their phones.

Interviewer: Is it common that students have access to the Internet on their phone?

Banaanimuusi: I only have wireless network access but it is hard to find an open network at school.

Panda96: But in our class most people borrow Matti's mobile phone to get access to the Internet.

Banaanimuusi: We just listened to music on YouTube with Matti, a moment ago. (Interview with two 15-year-old girls)

Our informal conversation at the beginning of a visual arts lesson shows that the school space and presence in the space are duplicated. Students are present at the same time in the classroom and on Facebook, even during lessons. Most of the students, however, did not have access to the Internet via their mobile phones. According to the EU Kids Online survey, 36% of Finnish 9- to 16-year-old children use the Internet by mobile phone or with other mobile devices (Livingstone et al., 2011a, p. 23). If someone has access by phone, others can borrow his or her device. In these cases, mobile phones are communal and "public" devices, similar to phones in rural areas, for example in some parts of Africa (see, e.g., Bruijn, Nymanjoh, & Brinkman, 2009).

Fourteen-year-old Matti had a 3G phone and access to the Internet. He shared his phone with his classmates who wanted to check what was happening on Facebook or do something else on the Internet. I asked him what he thinks about lending his phone. "Well, it is quite funny when all of them ask me but usually I lend it for a moment," Matti answered. He thought that he lends his phone "a few times" a day, "a quarter of an hour altogether." Matti said that usually teachers do not notice it, but on a few occasions they have intervened.

Social networking is a part of everyday life at school and unofficial school space. Mobile devices are not the only way to go online; school computers also have access to the Internet and social networking sites. A peculiar mixture of schoolwork and social networking was especially present when students studied

---

1 These girls wanted to use the nicknames Panda96 and Banaanimuusi during the interview.

in the school's computer lab. For example, students did their schoolwork on the front page of the school computers while simultaneously making a few short visits to Facebook and to other social networking sites, especially if the teachers could not see them. Some of the students used Facebook as a music device while they were working in the class. Facebook and especially YouTube offer opportunities to listen to music for free.

When I asked students what their favorite media practices are, many listed visiting Facebook or other social networking sites. They argued that Facebook is a multisided platform where they can chat with friends, share images, and play games. Farmville was one of the most popular games when I did my survey in August 2010. From the perspective of school, it was rather problematic because students wanted to take care of their farm during the school day. This was one clear tactical practice that broke the official school space and made room for out-of-school practices. But it was not easy to find an available computer at school for playing Farmville. Some students even took into account that because it is not probable to find a free computer during the school day and visit Farmville they had to plan their harvest in advance. Janne, 15, explained:

> I thought that I do not have any chance at school to play Farmville and that's why I planned that I put something to grow about eight hours, so that I can harvest it after the school...but that was messy. I had to stop playing because sometimes I had to pick vegetables at 11pm or 1am and I thought that I do not want this anymore. (Interview with Janne, 15-year-old, male)

Janne and other students said that Farmville and Facebook as well are really hooking them. Teachers sometimes had to solve some conflicts that started on Facebook when students shared improper images or started arguments with other students. The principal said that some students had even changed school due to cyberbullying on the social networking sites.

For teachers, cyberbullying was a new phenomenon that was not easy to solve because the origin of the problems were outside the physical school space. Even though Facebook is not under the control of teachers or any adults, it is usually related to school. I use the term *school community-based practice* to describe social networking and other media practices related to the school community. With this I mean practices where students share photos taken at school, write a weblog related to the school and classmates, play video games with schoolmates, start a chat about school issues, have school-related groups on social networking sites, and so forth. Students may, for example, have groups under the name of a school teacher or groups for a particular class. For example, I observed students on IRC-Gallery who created a group bearing the name of their Swedish textbook.

School community with classmates and friends is an important part of the lives of youth. A great part of everyday communication is done with school friends and about the school. Social networking is not an exception. Here are some status updates by a 14-year-old girl on Facebook:

Status update 1: Oh no, I hate physics. I just don't understand anything.

Status update 2: Bye now, I go to read for an exam.

Status update 3: We still have not solved the problem of "kenä" and "kenenä."

The last status update was a comment about Finnish grammar with a local dialect. This shows that students can use social networking sites to ask questions about schoolwork or discuss school-related issues. For example, one of the students made a weblog entry where he commented on a discussion in the school class:

> Today we had a very interesting discussion [at the school] about a wonderful book *Ihana meri*.[2] The discussion changed from the book to the issue if a society creates too narrow models for women. My opinion is that media and society make very tight models what a "good woman" is. Magazines are full of miracles of photoshopping and weight loss diets, and top actors and models in television are part of it. The book, *Ihana meri*, was about anorexia and it discussed about these faults.
>
> But girls in our discussion argued that "only some stupid people" have these kinds of problems. They said that people who do not have a life or friends are victims of what they read in magazines and other media. This is partly true. But the girls commented that people who are good at school belong to the same "no life" group, and these people suffer from the models of women. In other words: you are out of your mind if you are good at school.
>
> This sounds silly, but I would like to have comments, am I mistaken or is there something odd with the girls' comments? (Weblog by 14-year-old student, male)

Discussions at the school are part of formal learning but there is usually too little time for 30 students to make arguments and no real way to continue or express their opinions when class is over. Social networking offers an opportunity for the students to continue the discussion and to have a voice and a public opinion. On social networking sites, opinions are no longer relegated to school business but are part of public discussion even as they are also part of a discussion in the school community. This is an example of permeable boundaries of local and global and public and private. Arjun Appadurai and Carol Breckenridge (1995) used the term *networked publics* to describe the public culture of the Internet. It means public spheres and active participation in the production of culture. Networked publics facilitate the development of public identities and peer-to-peer communication on the Internet.

In networked publics, young people act with their closest friends, local communities, and a global audience at the same time. Community-based practices on the Internet mean that the audience is not necessarily intended to be global, such as students are discussing local school issues. Rather, communal activities are a continuum of existing networks in school and everyday life. In this case, social networking strengthens and makes visible existing networks (cf. boyd & Ellison, 2007).

---

2 This book, written by Finnish writer Kira Poutanen, was awarded the Finlandia Junior Prize in 2001.

· 4 ·

CREATIVE MEDIA PRODUCTION

3-Cs Model of Media Literacy

In their book, *Media Literacy in Schools,* Andrew Burn and James Durran (2007) introduced their 3-Cs model of media literacy. Key aspects of the model are three social functions: *cultural, creative,* and *critical.* Burn and Durran argued that media literacy has all these functions. First, the *cultural function* refers to the media and popular culture that surrounds us but also to cultural practices like clothing, piercing, and skateboarding or whatever in a special place and time is culturally present and practiced. They especially emphasized cultural identity as an important cultural practice of media literacy: "We...see engagement with the media as part of wider cultural complexes of taste, pleasure and critical engagement, in which social identities are built and negotiated" (pp. 11–12). Using and producing media is partly an identity work where people represent themselves as well as their own tastes, interests, pleasures, knowledge, and expertise. In a way, identity is negotiated between the individual and culture and distributed throughout the culture and other people. Media culture, with different media products, habits, and styles, is a place for this kind of negotiation and distribution. In a way, people are the production of media culture that shapes and modifies them in the process of media consumption.

But at the same time, people have—at least to some extent—an agency and control over their identity and cultural circumstances. This leads to the second function of media literacy: the *creative function*. Burn and Durran (2007, p. 13) pointed out that in recent years, there has been a shift in emphasis from traditional critical dimensions of media literacy to *production* as a creative dimension of media literacy. They indicated this change by placing the creative function before the critical function. I add that this change is at least a small step toward participatory culture, where people are said to be active producers of the culture and are able to share and distribute their media productions in public spheres like the Internet and YouTube.

For Burn and Durran (2007), creativity means representation, communication, and the composition of coherent media texts. They referred to Lev Vygotsky's idea of playful activity, where "children learn the meaning of symbolic substitution through the manipulation of physical objects" (p. 13). A classical example of Vygotsky is a broomstick that represents a horse in children's play. In the context of media culture and media literacy, creativity is the capability of producing something valuable and worthwhile by using media devices and texts, by remixing, rewriting, and making media.

For David Gauntlett (2011), creativity means making things, a practice of craft, which includes making things online, such as creating, sharing, and publishing videos on YouTube or writing and publishing blogs on the Internet:

> Everyday creativity refers to a process which brings together at least one active human mind, and the material or digital world, in the activity of making something. The activity has not been done in this way by this person (or these people) before. The process may arouse various emotions, such as excitement and frustration, but most especially a feeling of joy. When witnessing and appreciating the output, people may sense the presence of the maker, and recognize those feelings. (p. 76)

An important aspect of creativity is the process of making and joy, some kind of "flow," that Mihaly Csikszentmihalyi wrote about in his books. Gauntlett (2011) celebrates the Internet and Web 2.0 as a new way to show creativity and share works globally. But there are also many counter voices. For example, Andrew Keen (2007) takes a critical point of view regarding the increased media production of average people. He suggests that the Internet has opened a free publishing channel to "millions and millions of exuberant monkeys" (Introduction, para. 9). He referred to Adolf Huxley's "infinite monkey theorem": "If you provide infinite monkeys with infinite typewriters, some monkey somewhere will eventually create a masterpiece" (Introduction, para. 7). According to Keen, instead of typewriters and masterpieces, amateurs of media production in the age of Web 2.0 are "creating an endless digital forest of mediocrity" (Introduction, para. 9). If we look at the World Wide Web today, this argument is hard to deny totally. In the digital world, the publishing threshold is low even for young people. But in most cases, their goal is not at all to make masterpieces but to have fun together and collaborate with each other. An important part of creativity is collaborating and connecting with other people who share the same interests.

Buckingham (2003b, p. 127) especially criticized the romantic concept of creativity, where creativity is seen in individualistic terms as the emanation of some inner or authentic self and spontaneous feeling. The classical notion of the "genius" emphasizes individual creativity out of nothing, pure self-expression and outsiderness. But this is not the concept of creativity as the function of media literacy. Rather, media literacy calls for the collaborative function of creative production. Richard Sennett (2009, pp. 21–22) recognized the Homeric Hymn to the master god of craftsmen, Hephaestus, as one of the first celebrations of creativity and creative craftsmen. The term for the craftsman in this hymn is *demioergos*, which is a compound made up of *demios* (meaning public) and *ergon* (meaning productive). The creative craftsman has little to do with personal genius in the modern sense, but more with public life, where skilled people participated in society by making things like pots and also being doctors, singers, and heralds, "who served in ancient times as news broadcasters" (p. 22).

The creative dimension of media literacy in the age of digital technologies also means a capacity to participate in society and cultural practices as well as articulate "news" as personal voice. However, creativity and making things are not enough without other functions. Burn and Durran (2007) articulated that the third social function of media literacy, the *critical function*, is also important. Buckingham (2003b, p. 313) has argued that media education is a part of the modernist project. It has gone hand in hand with the classical idea of critical consumer and the cultivation of rational and critical thinking:

> We [media educators] want to give our students the critical knowledge and the analytical tools that we believe will empower them, and thereby enable them to function as autonomous, rational social agents. Our aim is to produce critical consumers, even if what we mean by the term *critical* in this context is often ill defined. (Buckingham, 2003b, p. 313)

The role of education in media literacy is often seen from this perspective, familiar to cultural studies, especially in Great Britain. In the United States, critical media literacy is built both on cultural studies and critical pedagogy (Share, 2009, p. 3). From this point of view, critical media literacy is tied to the ideas of radical democracy and is concerned with developing skills and abilities that enhance democratization and civic participation.

At schools, for example at the Maple School, media education was seen by teachers especially from this critical perspective, as an opportunity to analyze critically media text, especially advertisements and journalistic articles. I return to this issue in Chapter 5.

For students, all these aforementioned functions are part of their media literacies, and what they bring to the unofficial school spaces is especially the creative function of media literacies. Many students that I interviewed told me that they make media, produce some kinds of media texts, and share them with others on the Internet. For example, 52% of students said that they publish and share photographs every month or more often. The sharing of all other media materials was not as common: 18% published drawings or paintings every month or more often, 16% blogged, 11% shared video content, and 8% wrote fan fiction (Figure 4.1).

By publishing and sharing their creations young people are making sense of their own identities and their place within broader communities. They mix different kinds of media contents and participate in social networks and productions. It is easy to recognize that modifications, remixings, and circulations of photos, videos, and music constitute the current media ecology in and outside the school. Remixing practices are blurring the boundary between personal communication and creative production and also passive consumption and creative production. The youth reworking and remixing derivative works can be seen as a form of literacy, where kids are creating personally meaningful and cultural works.

Figure 4.1: Media Production and Publishing One's Own Material (n = 305)

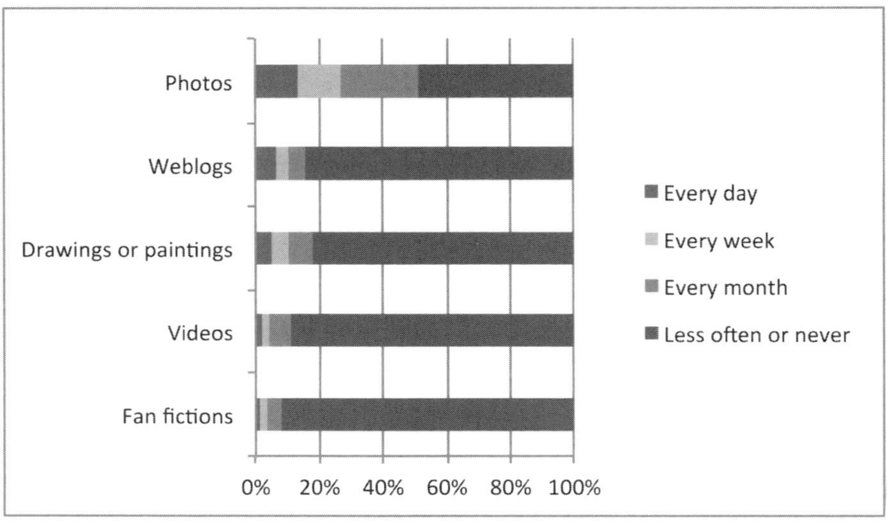

From the perspective of the social learning theory, this kind of participation is important. It does not consist of "local events of engagement in certain activities with certain people" but more of an encompassing process of being an active participant in the *practices* of social communities and constructing *identity* in relation to these communities (Wenger, 2008, p. 4). Such participation shapes what people do, what they are, and how they interpret what they do and are.

Youth Media Production

Making media has always motivated young people both inside and outside of school. Buckingham (2003b) wrote about his own school experiences as the person responsible for administering the media facilities in a large secondary school. He said that students were especially motivated to take their own photographs and to shoot and edit their own videos: "I can still recall some of the elaborate semiparodies of teenage magazines, 'youth programmes' and investigative documentaries that they produced" (p. 123). Since these kinds of school experiences, the media productions of young people have changed, especially due to digital media technology. As I mentioned before, mobile phones and smart phones that include still and video cameras serve as applications for digital editing. The majority of young people carry portable digital devices with them wherever they go and record everyday life, even at school.

Before the Internet, publishing one's own material in the public sphere was almost impossible. In the beginning of his book, *We-Think,* Charles Leadbeater (2009) described a fictive story of a young man who wants to publish a short clip of his guitar playing on TV. The young man, Funtwo, goes to see a controller of the TV channel in order to get a contract for his TV show. Here is how the conversation goes:

> *Controller*: So you want to make a video, what's it about?
>
> *Funtwo*: Of myself playing the electric guitar.
>
> *Controller*: I see—and you are planning to make this video in a studio with sound and lights?
>
> *Funtwo*: No I thought I could just use my bedroom. It's quiet and sunny.
>
> *Controller*: And if you are playing who will be shooting the video, where are the director, sound recordist and camera man…?
>
> *Funtwo*:…I thought I could just set up a video camera on a tripod and point it at myself. It's fairly simple.
>
> *Controller*: Do you have *any* experience directing or performing?
>
> *Funtwo*: None.

Controller: And how long do you intend this "video" to be?

Funtwo: Quite short, I only want about 5 minutes, 21 seconds.

Controller: Ah, that's a problem you see because the shortest programmes we do are really about 30 minutes. We'd really need something a bit longer. Do you have an idea when this video would be ready for transmission?

Funtwo: Well if I make it this afternoon I was hoping we could show it later this evening.

Controller: Ha ha, how charming. Let me explain the facts of life to you. You see, our schedules are full for the next nine months. Realistically we could not show anything for another year, at the earliest. And we deal in very large audience here, which is why we'd really need a celebrity angle, a well-known presenter and a format—a bit like *Ready Steady Guitar* or *Strictly Guitar*. Sorry. Final question: I don't want to sound unduly sceptical, because I am sure that you are really very talented, Mr Two but how many people do you think will want to watch you playing your guitar?

Funtwo: Oh about 51 million.

Controller: Ha ha, very droll. (Leadbeater, 2009, pp. xvi–xvii)

In fact, Funtwo is a Korean self-taught guitarist, Jeon-Hyun Lim, who has uploaded videos of his guitar playing to YouTube. When Leadbeater wrote his book, Funtwo had 51 million viewers. And what do we see in the video? The young boy on the edge of his rumpled bed, framed by the sunlight coming in through his window, head bowed, playing Pachelbel's baroque Canon in D Major with his guitar. The video has thousands of comments, such as "You are really a guitar hero," "Awesome," and of course some that consider the video to be a fake. Leadbeater's (2009) notion is that all of this could be otherwise. We can imagine what would have happened if YouTube or other digital publishing channels did not exist, and Funtwo had taken a different route to find his audience and gone, for example, to the British Broadcasting Company and asked whether he could make a video for TV. As we see from the imaginative discussion, it would be difficult to find "TV-time" for a short, homemade video.

What has changed during the time of YouTube and other digital publishing channels is that people now have an opportunity to upload and publish their media productions without media companies and controllers. Digital media opens new avenues for young people to create and share media without many hierarchical organizations. All kinds of "online content creation" are rapidly growing and have created a new kind of media ecology. YouTube and other social networking sites like MySpace and deviantArt enable young people to post, share, and discuss media works.

Youth media production is not only a form of creative media making, but it provides young people with "a variety of intellectual, aesthetic, social, and technical knowledge and skills they need to use, comprehend, and evaluate media," as JoEllen Fisherkeller (2011, p. 2) pointed out. In this way, youth media production is part of media education, especially when it encourages young people to participate in production practices that help them to be powerful members of society as active citizens, students, artists, or activists.

During my ethnography, I observed that some students use digital publishing channels actively, but it would be overstating to say that the whole generation of digital natives is keen to produce and publish their own creations. As we can see from Figure 4.1, the majority of students are not publishing anything on the Internet, or they publish just their photographs on Facebook and other social networking sites. But this does not mean that if they do not publish anything, they do not make media. Some of the students said in the interviews that they certainly make media, but they only share their products with their closest friends. For example, 14-year-old Janne said that he filmed short documentaries and "collages" with his pocket camera, and he had over 20 edited videos, but he did not want to publish any of them. He just liked to watch videos with his friends.

In this chapter, I discuss students' creative media practices both inside and outside of school. But I argue that all creative practices have some relation to school. I distinguish three different practices in relation to school: (a) school community-based practices, (b) out-of-school practices, and (c) curriculum-based practices.

By school community-based practices, I mean mostly friendship-driven media production in informal settings in unofficial school spaces. School is a meaningful space where peers meet each other daily. One of their activities, both inside and outside of school, is peer-to-peer media production using different devices such as stills, video cameras, and mobile phones. A school is a common arena for taking photos of friends and recording songs and other audio in different contexts. Activities of this kind are common in unofficial school spaces such as corridors and schoolyards, but in some cases, they take place in classrooms as well. Photographs and videos are shot without any planning and preproduction as a part of young people's social life and communication. Sometimes social networking, for example on Facebook, IRC-Gallery, and other social networking sites, is part of school community-based practice, for example, when students share photos taken at school, chat about school issues, form school-related groups, and so forth.

By out-of-school practices, I mean media making outside of school in informal learning contexts. Young people, of course, have different motives for their media production. At Maple School, many students worked with media and published their material just for fun and with friends.

Curriculum-based practices mean media production under the guidance of a teacher, related to the school curriculum and in official school space. Media production has specific goals and schedules and is supposed to take place in school during lessons. But this is not always the case. Students usually worked when they weren't in school, especially when they were strongly motivated and the timetable was tight.

Weblogs as School Community-Based Practices

One example of a school community-based creative media practice in my study is a video blog (vlog) produced and published by 14-year-old Gomi. Gomi was interested in movies and video making and planned to enter the movie business. In spring 2010, she launched a vlog about her everyday life in which school and school friends played a large part. In her video clips, she was just hanging out with friends at home, in the city and at school. Gomi and her friends had fun in their video clips: they played guitar, sang, danced, made food, watched movies, ate snacks, celebrated birthdays, and went shopping. At school, Gomi's video clips presented students' artwork during art class and drawings on the blackboard and showed students eating in the school canteen, hanging out, running in the corridors, and so forth.

Gomi's vlog presents everyday life inside and outside of school as a "carnival" in a Bakhtinian sense, in which the official school space, in particular, is turned upside down. Only young people are shown at school and they are embodied in the space in ways totally different from formal learning settings. The school space is represented as the students' own space without adult regulation and supervision. In official school spaces and traditional schooling, an embodied teacher rules the spatiotemporal product of the space and organizes schooling as a particular kind of activity (cf. Leander, 2007, p. 27). In Gomi's videos, the spatiotemporal structure is different, and here young people rule the space. A similar kind of unofficial space-time structure was also seen in some videos made in the mother tongue class in formal learning settings. Students took over the school space: they were dancing on the school desks and shooting videos in places where they usually had no access—such as the upper secondary school's gym—and beyond the boundaries of the schoolyard.

Video production is a good example of making space more homelike, especially because of the different bodily relation to the school space. Traditional schooling requires that students sit still most of the time in classrooms and apply themselves to the assignments at hand. Making videos seems to change this kind of bodily structure. Making videos is a bodily movement and surpasses the borders of the school. It goes out of the official time and space of school. Vivian Sobchack (1992) wrote that the nature of the film is that

> it uses *modes of embodied existence* (seeing, hearing, physical and reflective movement) as the vehicle, the "stuff," the substance of language. It also uses the *structures of direct experience* (the "centering" and bodily situating of existence in relation to the world of objects and others) as the basis for the structures of its language. (p. 5)

Video-making transformed Gomi's embodied existence in the school. Her vlog in a sense showed the school space as an unofficial space, not only as a learning environment but also as a social environment, which is, of course, what the school always is from the perspective of young people.

Above all, videos by Gomi are just for fun, connecting friends and recording and sharing everyday life. Creativity here means making things without thinking about how good videos are. Of importance for Gomi is the school community, her friends and online community, which are almost the same: the school community and especially Gomi's classmates are part of the video production.

Patricia Lange (2009) called these kinds of videos "videos of affinity" and "videos of affiliation." She highlighted the communicative and participatory dimensions of videos of affinity. Lange wrote that these kinds of videos are not necessarily interesting, original, or creative from the perspective of those who are not part of the creator's network and community. The content of videos of affinity is usually stereotypical, spontaneous, and contains several inside jokes that are not understandable to a general audience. Instead, "such videos are often communicative attempts to negotiate attention from other people to maintain ongoing connections or relationships" (p. 73). In some videos, Gomi even begs for comments from her friends: "comments please, now," she "asks" with some paper plates at the end of a video shot at the school. And she gets some comments on her blog: "You have to continue. Upload fun and different videos so we can enjoy watching them."

Affinity here means feelings of connection between people. Lange (2009) referred to anthropologist Bonnie Nardi, who has studied instant messaging in the workplace. Nardi noted that people often exchange short messages like "hi" without any other purpose than connecting with other people in the workplace. They created the affinity framework that was "crucial for keeping a 'labile' field of communication open in ways that later facilitated exchange of substantive business information" (p. 73). In the same way, Gomi's and other students' videos keep the communication channel open without any special film aesthetics. Videos look just as they are: homemade, simple clips with little or no postproduction. Lange argued that this kind of minimal aesthetics may promote more democratic uptake of filming and mediated self-expression, because videos are not dominated by standardized aesthetics or ideological discourses of filmmaking.

Dear Diary: A Day at School

In the aforementioned video by Gomi, she opens a school space for an audience of her vlog. In the beginning of the video, we can hear a male voice saying "dear diary." The name of the video clip, "Good Tuesday," also gives the impression of a diary, a regular day at school. The beginning of the video seems to be shot by a mobile phone: the image is small and not a regular widescreen picture. Gomi sits in a classroom at her school desk and explains that the teacher, who is coming in, is in a bad mood and "he is going to kill us all." There is no explanation of the reason for the teacher's bad mood and we do not see the teacher at all. The next scene after a transition is from the visual arts class and again without the teacher. Gomi and one of her friends are laughing and wrestling, writing on the chalkboard and just hanging out in the corridors. After this scene, Gomi is sitting at home (supposedly) and showing some paper notes: "comments, please," "now." This scene seems to be shot by a better video camera.

The length of the video is only 1 minute, 43 seconds. It does not have any special content or a message, but like in some other videos, it shows the school space as an "affinity space," as a space where Gomi and her friends and networks share and connect together. The "story" of an angry teacher is common anywhere, such as when old schoolmates meet and share school memories. Relations with the teachers and the characteristics of the teachers connect people together with the same affinities and emotions.

The use and presence of the body also promotes affinity (Lange, 2009, p. 78). In Gomi's video, this is shown by wrestling and walking in the corridors. In her video, Gomi used images that create a personal and communicative relation to the viewer: she talks to the camera and uses close-ups and jittery images, which are an index of spontaneous and personal interaction (see Lange, 2009, p. 78). In other videos, the same kind of spontaneity is present as well.

The body as an index that establishes affinity and breaks the official school space is also present in several ways in other videos by other students. For example, communal eating and drinking in a school canteen or café out of school, as well as dancing and singing together, promote affinity. These are also classical modes of a focal practice.

Focal practice or *focal thing* are terms used by the philosopher Albert Borgmann, which refer to practices and things that gather people in meaningful and social activities that have value for their own sake. Borgmann (1984) defined a focal practice as

> the resolute and regular dedication to a focal thing. It sponsors discipline and skill which are exercised in a unity of achievement and enjoyment, of mind, body, and the world, of myself and others, and in a social union. (p. 219)

A meal with friends is a good example of a focal practice, where people gather together and have an informal conversation. Each person is present in the same context sharing food and the joy of the common practice.

Gomi's community-based vlog is an example of how creative media production may strengthen school community, especially from the perspective of unofficial school space. A community and good "team spirit" have always played an important role in schooling, but usually they take place in official school space that is marked with school celebrations and sports meetings. In this case, school community is not built from the top down by teachers and school administrators but by students themselves. Video production is a technically important way to share feelings and meanings because it is itself bodily medium and "uses the *structures of direct experience*" (Sobchack, 1992, p. 5).

From this perspective, school community-based media production, whether it be making videos, social networking, blogging, and so forth, has an important role in school-related identity work and social cohesion, especially when it opens a new social space to the school, a space that we could call heterotopia, space with alternative social relations. At the same time, as identity work and creative practice, it is related to media literacies and skillful practices to create new cultural meanings.

Having Some Fun with Friends

We must keep in mind, however, that video production does not have a monopoly on the communal and focal practice. Students used other media as well, although not so much in relation to the school community. For example, Eve published photographs, drawings, and fan fiction in the deviantArt online community with her out-of-school friends. She described to me an episode of her media practices with her friends as "messing around with friends." Eve and three of her friends were on the computer and they decided to draw some "random" pictures, although they had a common interest in the manga series entitled *Axis Powers Hetalia* by Hidero.

Eve, 14, explained in an interview with me that this *Axis Powers Hetalia* "tells about the history and relations of different countries in II World War and these countries are presented as human beings." As human beings, they have human feelings and desires and II World War provides an interesting context to different kinds of relations and visual metaphors. Eve decided to draw Italy but in an obscure way: "Usually I draw Italy in a way that it shouts 'pasta' but now for some reason I draw it to shout 'swede pureé.'" Other friends drew some countries as well and also in a different way: "This one friend of mine drew Latvia that has this 'kolkol'-aura when normally it is that 'sob,' scared and shivering, because it has had traumas from Russia.... We really had fun."

Using a computer at home, Eve published the pictures on deviantArt and her friends had fun leaving comments. From the eyes of outsiders, it is very difficult to find the meaning of this "random drawing" and see it as a media production in the traditional sense. From my point of view, young people's media production includes all kinds of material and digital media making, from creating pictures to blog writing and filming. Media production is the making of media texts and representations and the remixing of these texts. The manga series, for example, provides fans with a great opportunity to make stories and develop a new kind of relation between countries as human beings. This kind of "textual poaching" and remixing have been of interest to many scholars, like Henry Jenkins (1992), Colin Lankshear and Michelle Knobel (2006), Rebecca Black (2008), and Lawrence Lessig (2008).

Fans write new stories, make home videos, reedit original TV series and movies, make music videos using film clips of the original products, and do other creative reworking of existing materials. A fan is not a passive viewer of a particular program but an active translator of that viewing "into some kind of cultural activity, by sharing feelings and thoughts about the program content with friends, by joining a 'community' of other fans who share common interests" (Jenkins, 2006b, p. 41). Especially important is that for a fan, "consumption sparks production, reading generates writing," and so forth (Jenkins, 2006b, p. 41). This activity is usually shared in fan communities' out-of-school practices.

Remixing as Literacy Practice

This kind of out-of-school media production has, in many cases, relations to popular culture and its characters and stories. The most common way to create one's own products is to combine elements of existing media productions. This way of producing media is remixing; it is the "practice of taking cultural artifacts and combining and manipulating them into a new kind of creative blend," as Lankshear and Knobel argued (2011, p. 311). For them, remixing is a key concept for thinking of new literacies.

One of the most well-known advocates of remixing culture is Lawrence Lessig (2008), who claimed that at the broad level, culture can be understood in terms of remix. The culture is a construction of ideas, artifacts, artworks, and other things that people have mixed together. Lessig (p. 82) pointed out that, in fact, writing with words and sentences is remixing when we quote, incorporate, or rearrange what others have said, just as I am doing now while writing this book. In the broadest sense, using language is remixing: we remix when we take a word and integrate it into sentences and speech. Writing is a most obvious practice of remixing, but we do not usually recognize it. Language is common; everybody can use it for free but the use of published photos by others is usually not possible without permission. There are copyright laws and other restrictions that do not allow us to remix whatever we want. But this does not prevent young people from "mixing" music, images, photos, or other "media texts" in order to create new meanings. Eve and her friends used the images from *Axis Powers Hetalia* and made new stories. In fact, one important part of fandom is making remixes.

Lankshear and Knobel (2011) took from Lessig (2008) the idea that remixing is a new way to write, some kind of extension of writing. It includes "practices of producing, exchanging and negotiating digitally remixed texts, which may employ a single medium or may be multimedia remixes" (Lankshear & Knobel, 2011, p. 313). For Lessig (2008), remixing culture is Read/Write (RW) culture, where people not only consume texts but create and recreate culture by making texts and media. The opposite of RW culture is Read/Only (RO) culture: "a culture less practiced in performance, or amateur creativity, and more comfortable (think: couch) with simple consumption" (p. 28). Remixing is RW or everyday creativity, nowadays reading and writing especially with digital media. Lankshear and Knobel provided many examples of creative remixing of culture and practices with different media: photoshopping remixes; music and music video remixes; machinima remixes; moving image remixes; original manga and anime fan art; television, movie, and book remixes; and serviceware mashups.

Eve and her friends and some other students in the Maple School used manga and anime fan art as a remixing practice. Six percent of students said that they visit the deviantArt web community every day, 4% every week, and 6% once a month or less frequently. Some said that they make anime videos and some used to visit other anime and manga communities. Teachers at the school recognized the popularity of manga and anime practices, but they did not utilize it in the classroom. A visual arts teacher told me that he is very interested in what students are doing out of school and has asked students to present what they have made to him. Some students even from the upper secondary school contacted him in order to tell him that they have, for example, new cartoons on their website. "But in the class...I do not utilize it at all.... What they do at home is their own thing." Other teachers agreed, and one mother tongue teacher stated the following: "Not many [students] want to tell these things to a teacher.... The Internet and other stuff are their own area...very much an intimate thing.... " Some students also said that they do not want to bring their creative practices into official school spaces.

The gap between youth culture and school culture is especially evident when students make this distinction. Making connections or bridges between these "two cultures" is therefore not a simple task. There are generational, curricular, and cultural gaps, but there is also the question of a personal life and interests, which cannot be shared with adults or other people who do not share same interests and which belong to the different communities out of school.

But even though some practices are out-of-school practices, this does not mean that they are out-of-learning practices. Remixing, for example, is interesting from the perspective of media and literacy education. Lankshear and Knobel argued (2011, pp. 327–328) that remixing practices enhance and require three kinds of literacy practices, which they called *technical, discourse,* and *evaluative* dimensions of literacy.

The *technical* dimension of literacy practice "involves knowing one's way around the processes and tools for encoding the meaning one seeks to articulate" (Lankshear & Knobel, 2011, p. 328). Remixings are uploaded and downloaded to and from different communities on the Internet, and made by using different digital devices and software. Young people have to learn to use these devices and software in an appropriate way. The question is not only the technical skills but also an understanding of communicative dimensions of remixes and expression of the self. Remixing is not just copying preexisting media content. Henry Jenkins (2006a) has argued that "more and more literacy experts are recognizing that enacting, reciting, and appropriating elements from preexisting stories is a valuable and organic part of the process by which children develop cultural literacy" (pp. 177–178). I must emphasize that children and young people do not have any technical skills or literacy knowledge as in-born natural attributes. On the contrary, they learn them, sometimes in formal and sometimes in informal or non-formal settings. Mimi Ito and her research group (Ito et al., 2010) spoke about interest-driven learning, when the learning is driven by interests learners have. RW or remixing culture can offer interests, but it is not for everyone.

Another dimension of literacy practices is what Lankshear and Knobel (2011) described as *discourse*. The discourse dimension "involves bringing cultural knowledge to bear on the tasks or purposes of the practice in which one is engaged—how to mobilize and co-ordinate the meaning elements" (p. 328). Making remixes requires cultural and sometimes ideological and political knowledge of media contents. In fan studies, a fandom is frequently understood as a site of countercultural activism and political practice that turns meanings of mainstream popular culture upside down. For example, one common genre of fan fiction is *slash*, where homosexual relationships are at the center of the stories and women, especially, can explore and negotiate their sexuality by reading and writing their desires (Busse, 2008). Fan fiction has also appealed to minorities and individuals on the cultural margins, but it is too strong to say that it is a genre of "pure" resistance, as Fiske (1992) has pointed out (see Derecho, 2008).

Nevertheless, fan fiction as an example of remixing requires knowledge of discourses, for example what rules, norms, and criteria apply to literacy practice, "what kinds of shared meanings circulate within members of the practice and the kinds of signs, symbols, sounds (meaning tokens) that bespeak these, and how different co-ordinations out of these can convey different meanings" (Lankshear & Knobel, 2011, p. 328).

The third dimension of literacy practice is called *evaluative*. It means knowing "how to enhance or improve the practice in order for it to better fulfill the interests of those who engage in it and who are impacted by it" (Lankshear & Knobel, 2011, p. 328). Evaluation is an important part of every learning process. As a dimension of literacy practice, it includes some concrete actions in order to enhance practice, for example by developing better guides for remixing or cultivating beta-reader activity in fan fiction communities.

Ola Erstad (2008) argued that remixing activities are an essential part of digital literacies that can change traditional book-based and teacher-driven schooling toward a situation where students create something new and not predefined (p. 178). He defined remixing as

> selecting, cutting, pasting, and combining semiotic resources into new digital and multimodal texts (bricolage), which is achieved by downloading and uploading files from different sources (Internet, iPod, DV-camera, digital camera or sound recording devices). (p. 186)

The idea of bricolage comes from the tradition of cultural studies, where scholars studied subcultures and their ways of adopting signs and symbols and placing them into a new context that erases or subverts the original meanings. Examples are mods[1] with their grubby parka anoraks and scooters or punks with swastikas and safety pins on their face. In the digital age, young people adopt digital content that they transfer to a new context by combining different media elements into new textual expressions. For Erstad (2008), this is a "fundamental change in the way we conceive reading and writing as cultural practices of meaning making" (p. 198). He spoke about trajectories of remixing where young people work by using different information sources and media modes, selecting them, combining them in different ways, creating new ensembles, and sharing them with others for new mashups and uses. As digital literacies, these kinds of practices are examples of moving away from a simplistic understanding of media and digital literacy as the skills operating in technology to a more complicated way of working with multimodal media texts. As Erstad noted, "Remixing encompasses many of these competencies, such as selecting, organizing, reflecting, evaluating, creating and communicating" (p. 198).

As we can see, digital literacies as remixing practices are comprised of different kinds of skills and competencies in new social contexts. The traditional definition of media literacy as "the ability to access, understand and create communications in a variety of contexts" is too simplistic and does not reach the diversity of digital reading and writing processes or RW culture.

Beta Reading in Fan Fiction Practices

The aforementioned dimensions of literacy are identifiable in remix practices as processes of text creation. One of the most interesting practices discussed from

1 Mods belong to the mod subculture that originated in England in the late 1950s and early-to-mid-1960s. One significant elements of mod subculture includes a special kind of fashion, like suits and military parka.

this perspective is beta reading in fandoms. The name *beta reading* comes from *beta testing* in computer programming. It is the practice of releasing a story to a fellow writer in the fan fiction community. Usually, the beta reader is a selected and trusted reader in an online fandom. The task of the beta reader is to read a draft of a story and give some feedback and suggestions for improvement of the story. Suggestions may apply to all aspects of the story, from grammar and spelling to narrative, structure, and characters. At the same time, beta readers help writers maintain personal participation with the community.

In the Maple School, 5% of the students reported that they are active beta readers, which is almost as many students who reported being authors of fan fiction (7%). One of the beta readers and authors, 15-year-old Uni, said that she started to write fan fiction in Grade 7 on the Finnish *Harry Potter* forum, but soon her interest changed from *Harry Potter* to J-rock (Japanese rock) and Korean pop. She reads a lot of fan fiction from *LiveJournal* and has written some stories for *Lost and Found*, which is a Finnish-speaking fan forum for J-rock fans. She has also been a beta reader on *Lost and Found*.

Uni told me about her practices with a writer for whom she had been a beta reader for about 2 years. She used to go through the text three times with the writer. The first time was a general reading, and the second time she would correct the writer's typos and spelling errors. The third time was a discussion with the writer via Microsotf Messenger. Uni said that some of the items they discussed were sentence structure, what kind of impression the writing was conveying, and the interpretation of the text. These discussions helped the author improve her stories. Uni insisted, however, that she did not change anything in those stories: "I always asked [the writer's] opinion and told her what is unclear." As Uni noted, the writers were often so good at writing that she did not need to correct the narrative or any other structures of the story. Uni also had a beta reader for her own stories, and she said that comments were, in most cases, positive and supportive: "I've never got comments that affected me badly or anything." The most important aspect of this process, especially from the perspective of learning, is positive feedback, which also expresses the beta reader's affiliation with the author's work.

The role of the beta reader is very interesting and special in terms of fan fiction and in relation to the text and readers. Angelina Karpovich (2008) wrote that the relationship between the beta reader and the text combines different elements at the same time. First, a beta reader is a reader who is the target audience of the text; second she is the critic and commentator who reviews the text, and third, she is the editor who ensures that the text fulfills certain requirements. This complex relationship differs from the work of a copy editor, especially because the beta reader "is a priori understood to be approaching the text as somebody who is entitled to be just as invested in the characters and their story as the author" (Introduction, para. 12). The reader and writer operate within the same frame of reference and share the same interests and emotional investments. And more than that, both of them operate as active creative producers and manipulators of meaning (Karpovich, 2008).

In a way, beta reading and fan fiction writing are parallel to school-based literacy practices such as composition, grammar, editing, and reviewing (Black, 2009). The interesting question that Rebecca Black (2009) raised is what is so engaging in these kinds of practices and informal learning sites that young people willingly devote hours participating in them? (p. 600). Answers to this question can be found by following different traces. From my point of view, the answer may be a spatial one: out-of-school literacy practices open new spaces where different mindsets, perspectives, and practices come together and give room for different people to use and enhance their own strengths and competences. These spaces value hybrid and multiple forms of meaning making; in schools, these are still more or less marginalized, but they are increasingly present because of the students' diverse media and literacy practices.

Tactical Literacy Practices and Textual Poaching

Henry Jenkins drew on the work of de Certeau and the concept of poaching in order to explain strategies of fans who appropriate texts and rewrite them in new contexts and circulation of meanings. For Jenkins, fans are a model of the type of textual poaching that de Certeau (1984, p. 175) explicated as being characterized by "advances and retreats, tactics and games played with the text," and cultural bricolage, through which text users and readers take existing texts for their own use appropriating and reappropriating them:

> Readers are travelers; they move across lands belonging to someone else, like nomads poaching their way across fields they did not write, despoiling the wealth of Egypt to enjoy it themselves. (de Certeau, 1984, p. 174)

Poaching challenges the very notion of literature and texts as a kind of private property controlled by textual producers. Poaching in a way robs, "misuses," and fragments a text and reassembles it according to poachers' experiences and own use of the text. De Certeau (1984) pointed out that texts in schools, for example, are controlled by teachers, but at the same time there are tactical practices, "the silent, transgressive, ironic or poetic activity of readers...who maintain their reserve in private and without the knowledge of the 'masters'" (p. 172). He wrote that a school child, for example, "scrawls and daubs on his schoolbooks; even if he is punished for this crime, he has made a space for himself and signs his existence as an author on it" (p. 31). In a similar way, David Barton (2007) pointed out alternative literacy practices in schools:

> In the classroom there is graffiti and doodling, names are carved on desks, secret notes are passed. Children read comics and circulate illicit material, they have their own books and magazines brought in from the outside. In addition the visual environment of notice boards, classroom displays and signs surround children. (p. 177)

Names carved on desks and scrawls on schoolbooks are examples of everyday tactical literacy practices at heterotopias, where students inhabit both the official and unofficial school space at the same time. This is even more obvious in the digital age, when students use their digital devices to write text messages or fan fiction or update their status on social networking sites.

These kinds of literacy practices are part of textual poaching and tactics where the "reader produces gardens that miniaturize and collate a world, like a Robinson Crusoe discovering an island" (de Certeau, 1894, p. 173). As I have noted, these literacy practices are tactical and certainly not part of official school space and traditional learning. Jenkins (2006b) wrote the following:

> Not surprisingly, someone who has just published her first online novel and gotten dozens of letters of comment finds it disappointing to return to the classroom where her work is going to be read only by the teacher and feedback may be very limited. Some teens confessed to smuggling drafts of stories to school in their textbooks and editing them during class; others sit around the lunch table talking plot and character issues with their classmates or try to work on the stories on the school computers until the librarians accuse them of wasting time. They can't wait for the school bell to ring so they can focus on their writing. (p. 184)

As a result, students try to create unofficial school spaces and find their way to new literacy practices inside the school. Of course, writing fan fiction at school is not a common activity shared by many students, but it is an example of the literacy practices that have brought out questions of learning in new, sometimes more participatory contexts.

Writing fan fiction may be an individual practice, but as part of a fan community and beta-reading processes, it is a more participatory practice with other fans. Fan cultures usually offer different kinds of processes, where people can share their skills and work together. One example is fan translation.

Fan translation is an unofficial translation of some production of popular culture to English or other languages, where people translate popular publications in their own language before a publishing house finds the time to do so. For example, in China, a group of about 60 university and secondary students translated *Harry Potter and the Deathly Hallows* into Chinese only 2 days after the global release of the original text and almost 3 months before the official translation in China (Xu Fan, 2007).

Learning in Affinity Spaces

For Lessig (2008, p. 77), education is an important aspect of remixing practices and another is community. Remixing practices and writing fan fiction typify interest-based learning, where participants are developing their skills together with others who share the same interests. Uni described fan fiction practices mostly in terms of interest-based and peer-to-peer learning. She mentioned that some of the authors had a long practice of writing, such as "some 20-year-old people who have been writing from the age of ten and have been in all courses and are studying literature history and other stuff." In this way, fandom is a kind of writers' academy, out of the formal learning context, but at the same time the activity calls forth interest in further studying. In the fandom, the entire community takes first responsibility for helping new writers enhance their skills.

This is a good example of the situated learning that James Paul Gee (2007) described in terms of *affinity spaces*. Affinity spaces are participants' own spaces with special peer-to-peer relations and learning settings without previously founded tight social bonds. Members of affinity space relate to each other, for example, in terms of a video game or in different fan fiction communities:

> An affinity space is a place (physical, virtual, or a mixture of the two) wherein people interact with each other, often at a distance (that is, not necessarily face-to-face, though face-to-face interactions can also be involved) primarily through shared practices or a common endeavor (which entails shared practices), and only secondarily through shared culture, gender, ethnicity, or face-to-face relationships. (Gee, 2007, p. 98)

People join affinity spaces due to their shared practices, interests, and common endeavors. For example, spaces for fan fiction are spaces of fans who share the same interests with other fans. Gee (2007, p. 78) wrote that the concept of space avoids the problems of the concept of community of practice that is used by Lave and Wenger. Etienne Wenger (2006) defined the concept of community of practice in the following way: "Communities of practice are groups of people who share a concern or a passion for something they do and learn how to do it better as they interact regularly" (n.p.). From this perspective, learning is looked at in terms of apprenticeship, when learners "apprentice" themselves to a group of people who share and improve a certain set of practices. However, Gee (2007, p. 78) criticized the concept of communities of practice. He suggested that a classroom cannot be a community, which consists of close-knit personal ties among students. A school or a classroom does not necessarily constitute a group of people as it is thought in the concept of community:

> Take a high-school science class. Johnny and Janie are both in the class. Janie is proactively attempting to engage with the science in the class, but Johnny is "playing the game" for a passing grade. Are they in the same community of practice or is Janie in a school science community of practice and Johnny in a "doing school" community of

> practice? What sense does it make to say all the students in this class are in some (one?) community of practice just because they are all contained by the same four walls? (Gee, 2007, p. 78)

Instead of communities, Gee (2007) emphasized spaces, which allow people to join together with shared practices and use multiple forms of participation, knowledge, and experience. In affinity spaces, all kinds of practices are valued, some of them more intensive, with some participants bringing specialized information into the community, and some of them are more extensive, with understanding of the site on a broader level and sharing the information outside (Black, 2009, p. 598; Gee, 2004, p. 87).

But as we have seen, these are spaces out of school. Jenkins (2006a, p. 184) suggested that affinity spaces cannot be duplicated by incorporating similar activities to the school. Young people need their own places where they can read, write, and practice textual poaching without the leadership hierarchy and grammar of schooling. Out-of-school media production has a different space-time structure from schooling. For example, students may work online with peers and concentrate on their assignments for hours without a break, spending their time in cyberspace without standardized organizational practices. For example, Santeri, a 14-year-old online forum role-player, spent plenty of time designing his avatars by describing all their characteristics:

> A good avatar is well designed ... and you can recognize an avatar that you designed in detail over many days ... I'm a perfectionist in terms of avatar design. I always begin by considering in what way they're badly designed. I would like to do a very good avatar that I wouldn't immediately begin to change or criticize in my mind. (Interview with Santeri, 14, male)

Santeri was a participant in a small, online forum role-play community of 30 peers. This community is a virtual world called Lyrath where every participant has avatars or characters that inhabit the virtual world. Participants do not make representations in the sense that large audiences are supposed to see them. They are playing together, and for an outsider it is even harder to understand what is happening without knowing the rules, history, and regime of the world.

Santeri said that an idea for an avatar might come for example in a book he has read, and from this point of view, it is also a practice of remixing. Santeri said that he tries to create an avatar that is totally different than himself:

> It is to take a track at how it is to be somebody else. I try to think what another person would do. Some gamers have more avatars that are like them. It is easier to play, but I like to try what it is like to be someone different. (Interview with Santeri)

Santeri had several avatars in the role-play, for example an elf called Loki. Santeri has written detailed characteristics of Loki, including its profession, age, gender, sexual orientation, residence, weapon, element, family, character, personal history, look, skills, weaknesses, friends, pets, diseases, and so forth. All the characteristics have to be in relation to other avatars of the play and of course to the whole world where the play is actualized and where the avatars live and meet each other.

In the role-play, every gamer creates avatars and writes a story of these avatars with others. Writing is collaborative work:

> *Interviewer:* How does it work?
>
> *Santeri:* It is a kind of collaboration, usually two people are writing together or we write in a bigger group. One person writes first, then another, and then the first one continues. The story is so much more vivid and funny to read.
>
> *Interviewer:* Is it some kind of continuation?
>
> *Santeri:* Yes, but who is playing writes from the perspective of his own avatar, and I write from the perspective of my own one. (Interview with Santeri)

The world of Lyrath has some rules. For example, a writer cannot do anything to another's avatar without permission. Some rules are for writing. A gamer has to write, for example, what an avatar thinks, feels, and perceives while using good language. Lyrath has a history, ecology, its own currency, lords, and so forth. There is even a map of Lyrath. Stories are located at places where avatars live and only the administrators of the game are able to change anything. For example, at the end of August 2010, administrators reported in the forum that a town called Mihras had been burned. This was an unpleasant incident for many of the gamers. They had to take this incident into account in their avatar's life. To burn the town seemed to be an effort to activate the game.

As de Certeau (1984) pointed out, poaching "is to constitute a secret scene, a place one can enter and leave when one wishes" (p. 173). Santeri and other participants have their own world with its own rules and organizations. They share practices with peers with the same affinity, interest, and endeavor. This practice has nothing to do with the school. As Santeri noted, "They [avatars he has made] live in their own world...they don't belong at school."

Avatars in different Internet communities and, for example, in the virtual world Second Life, are usually part of an identity play, "performances of you," and they are not only who you are but also who you perhaps want to be (Turkle, 2011). As Santeri said, he tries out "what it is to be someone different," to be someone else. He thinks that his online identity in a game has a life of its own and is separated from his real life. Virtual space exists in this way as a coherent social space of it's own with new rules and ways of being, as parallel to the unofficial school space and therefore separated from the official school space. It is another space, heterotopia for identity play. Trying to be something different than yourself may be challenging but also empowering. This is seen especially in fan fiction culture and storytelling in general, where writing offers positions to challenge, for example, dominant gender roles.

For Santeri, the most important aspect of participating in the world of Lyrath was self-expression through writing. He said that through role-play games he had learned to write better: "I'm not learning very much there, mostly it's just fun, but perhaps I'm a better writer now." In fact, by writing stories, Santeri not only improved his skills as a writer, but he also had to think of stereotypes of gender characteristics, narratives, and structures of the storytelling. He also shared his thoughts in another arena, his blog, where he wondered about different stereotypes of gays and different nationalities.

Fan fiction and role-playing practices are both examples of peer-based learning that is based on peer sociability and online communities out of school spaces. Ito and her colleagues (2010) argued that peer-based learning "relies on the context of reciprocity, in which kids feel they have a stake in self-expression as well as a stake in evaluating and giving feedback to another" (p. 22). This kind of reciprocity has a different dynamic than more hierarchical and authoritative learning relations.

My study indicates that because some students bring their out-of-school literacy and media practices to the school, they create heterotopical spaces to the school, spaces where they find new ways of learning and practicing. In a way, traditional structures of school "reject" new literacy and learning practices, but school literacies and out-of-school literacies intertwine together. The relation is complicated and students' media practices have different relations to the schooling and official school spaces, some more out of school, some more related to school community and curriculum-based teaching and learning.

Lifestyle Blogs and Photo Performances

Writing a lifestyle or fashion blog is an example of out-of-school practices, but it is interesting because blogging is one of the most popular creative media practices among young people and especially among young girls in Finland. I found almost 4,000 hits from the Finnish web portal blogilista.fi (blog list) using the keyword *lifestyle,* and over 3,800 hits with the keyword *muoti* (fashion).

I use the term *lifestyle blog* here because the girls in Maple School do not directly present fashion, but their own life, in which fashion plays a great role along with some other things like food, friends, school, and trips. Writing a lifestyle blog requires a variety of literacy practices from storytelling to making and editing photos and managing HTML language or the use of publishing editors. It also entwines an identity to literacy practices. In a way, a lifestyle blog is a literacy practice to build, control, and enact identity in a special way.

In the Maple School, 14% of the students reported that they publish a blog once a month or more often, while 6% reported doing so every day. Of all the bloggers, only four were boys. One of the bloggers from the school is 14-year-old Susanna, who started her lifestyle blog in 2007 when she was 12 years old. First, she was interested in fashion and started to read fashion and lifestyle blogs. Then she decided to have her own blog. In 2010, she started her second blog with 50 followers and many more readers. Tara Chittenden (2010, p. 514) pointed out that the notion of "follower" is interesting. Usually on the social networking sites, like on Facebook, the term *friend* is used, which denotes someone with which the person has a connection or close relationship. Instead, the term *follower* suggests a hierarchy, "almost a religiosity," in the relationship between the blogger and the follower (p. 514). Followers are people who wait, read, and comment on blog entries, and in this way, a blogger is an author with an explicit audience.

Thomas and Brown (2011) wrote that blogs generate the space for a collective to emerge. Blogs rest in part with the blogger,

> in part with the readers who leave comments; in part with those who link to, cite, reference, or respond to it; and in part with the readers, who may do nothing more than have their presence recorded by a web server. None of those events alone is sufficient for understanding the phenomenon; it is the combination of the active and passive (such as comments, ratings, and links) forms of participation that makes a blog or website successful. (Thomas & Brown, 2011, Learning in the Collective, para. 3)

As a learning environment, learning within this context emerges from the dynamics of this collective, where a blogger, with others, is facilitating the construction of a learning community. Blogging is not traditional writing to audiences and it is not a traditional literary form, rather it is a kind of expression between orality and literacy, perhaps a *secondary orality,* to use Walter Ong's term (see Dean, 2010, p. 48). The theory of secondary orality has its roots in the distinction between orality and literacy as developed, for example, by Marshall McLuhan (1962) and Ong. These authors have developed the idea that societies have been radically transformed by the adaption of literacy. Literacy allows information to be exchanged and shared without the need for a physical copresence and face-to-face relations. This leads to larger and diverse collectives with less personal connections. Electronic and digital media, instead, may create again personal connections and a sense of closeness, even among people who are distant in space and time. The notion of secondary orality refers to this virtual closeness in online groups. Primary orality refers to human expression without visualized words. Secondary orality "floods the world with sounded words (and other sounds) again. But it is the product of a writing and print culture" (Ong, 1994, p. 3). Perhaps we can also use the term *vernacular literacy* separated from institutional literacy (see Chapter 2).

Susanna mentioned that she blogs daily and sometimes more than once a day: "It is fun to write and make collages; take photos and modify them." The content of the blog includes mostly entries of the blogger's day and things she likes, especially music, food, and clothes. Images on her web site are a mix of photos, but mostly they are pictures she has taken of clothes, food, landscapes, buildings, personal items, and, of course, herself.

Susanna said that she has not shared her blog address with friends and classmates: "It is much more fun to write to totally unknown people than know that some friends read them." Susanna's lifestyle blog is an out-of-school practice and not in direct relation to the school community. However, her blog entries include comments about the school day and events at school. Blogging activity intertwines with school life as a part of meaning making. But the audience for a blog is not a preexisting school community, and it is not a one-to-one communication to friends. Jodi Dean (2010) suggested that a blog is a medium that enables content production to be potentially accessible to anyone who happens to find it: "Blogging opens possible encounters with the different and unexpected, whether in the form of the blogger's own reflection on what she posts or in the reactions of others" (p. 46). Perhaps that helps explain the popularity of blogging among young girls: blogging opens a new and exciting world where one can create unexpected connections to other people and an identity and body performance.

Secondary orality or vernacular literacy, which are somewhere in between orality and traditional literacy, includes attributes that are key features of online communication and, for example, blog entries. Dean (2010, pp. 48–49) highlighted the following attributes:

- Thoughts are combined and points are made in ways that are additive rather than supportive; differently put, people string syntactic elements together with "and" rather than with subordinate clauses.
- The elements of thoughts and ideas are aggregative rather than analytical (a contemporary example might be the slogans, clichés, and memes that catch on and stand-in for ideas and feelings that remain unexplored).
- Ideas and points are frequently repeated.
- Traditions are conserved (because little to nothing is written down, remembering is difficult; hence, not only do points need to be repeated but they need to be attributed to tradition).
- Ideas are understood in terms of their connection with actual experience, with the lifeworld, rather than abstractly or within a more general analytic field (*if I don't know anyone who has been the victim of homophobic violence, then homophobic violence must not be a problem*).
- Knowledge and ideas appear agonistically, polarized, as part of everyday struggle.

- Ideas are treated in terms of empathy and its lack: that is to say, in an immediate and participatory rather than a distanced fashion.

The practice of blogging, including networking with others and participating in daily discussions, seems to be more important than the content in blogs. Dean (2010) seemed to emphasize the idea of blogs as "oral communication in linear writing," where "I" is the center of communication: "I contribute to mark not simply the fact that I am here, that I have a view, but that I am engaging. I am participating in the construction and extension of a manner of being together" (p. 47). Zizi A. Papacharissi (2010, p. 144), for her part, characterized blogging as new narcissism but not in any pejorative meaning or in its pathological sense. Rather, "narcissism" here refers to the "introspection and self-absorption that takes place in blogs and similar spaces" (p. 145). It is also an expression of the more self-centered culture and privatization of the individual in modern society, as attributes that Dean highlighted indicates as well.

These perspectives seem to be the most common ways to characterize teens' writing activities. The perspective highlights blogging as *networked individualism*, a term used by Mesch and Talmud (2010). This term is adapted from sociologist Barry Wellman, who thinks that society has changed in modernity from a form of rather small homogeneous social groups to a form of individual basis groups. Networked individualism leads to interpersonal relationships around shared interests, common lifestyles, and short-time collaborations. Blogging can be seen as a mode of networked individualism and identity performance.

Ervin Goffman has called impression management the process of identity performance, where people give an image of themselves by clothes, speed, facial expressions, and other bodily impressions. danah boyd (2008) summarized this as follows:

> Impression management is part of a larger process where people seek to define a situation through their behavior. People seek to define social situations by using contextual cues from the environment around them. Social norms emerge out of situational definitions, as people learn to read cues from the environment and the people present to understand what is appropriate behavior. (p. 128)

Impression management requires social skills that we learn in everyday social situations. These skills are necessary when individuals are being socialized into a society and have "street credibility" in particular contexts. Impression management is also important on the Internet, where bodies are not visible in the same way as in off-line communication. However, the Internet allows more opportunities to construct an impression of the self and make identity performances. Agents on the World Wide Web can use text, images, video, and audio for identity play. But there are more dangers: the Internet remembers all and circulates information without the control of the sender (cf. boyd, 2008).

Blogs and social networking sites are channels to identity performance and publicity. boyd (2008) made an important point that publicity on the World Wide Web, what she called networked publics, has an important role in the construction of society: "Without publics, there is no coherent society. Publics are where norms are set and reinforced, where common ground is formed" (p. 137). Publics allow us to have collective experiences and also test our ideas, thoughts, and identities.

Chittenden (2010) argued that female teen blogs support the blogger's fluid notion of identity "as bloggers experiment with various looks, play with representations of themselves, and use various affinities with followers to build social capital" (p. 513). They create a blog as "a space for play," where they can conduct experiments with fashion and importantly gatekeep "access to their personal field of identity formation" (p. 513). Hille Koskela (2009) has pointed out that identity performance may take on political and empowering dimensions when people play an active role in the array of visual representations. Bloggers are able to decide for themselves about their body image on the web diary and control their information disclosure.

Angela Thomas (2007, p. 6) wrote that online communities show a cultural production of a new type of body. In these communities, the body is self-produced through words and images. It is a question of "the semiotics of identity," where an identity is a performance of fantasy and desire—"a pursuit of being and becoming the image of this desire" (p. 5). Online communities and one's own blog give a person the opportunity to experiment with identities and modify body images. Blogs, in particular, are good platforms to experiment with because they generally contain less storytelling than presentation of moments via an image, reaction, feeling, or event (Dean, 2010, p. 47).

It is also important to notice that activities on social networking sites are tightly connected to photo performances and identity play, that is, indeed, "the work of adolescence" (Turkle, 2011, p. 179). Photos on social networking sites are not only statements about who you are but who you want to be. Selecting a photo is often the very first thing to do on the social networking site. When 13-year-old Marjukka created her Facebook profile, she took plenty of time to discuss photos of herself with her friend. When she did not find a good photo from the computer's photo album, she decided to shoot a new one. She and her friend found a sunny place on the balcony for a photo session. After a "pose," as Marjukka said, she was ready to create her profile.

Commenting on photos is one of the most popular activities on social networking sites. People using Facebook write and rewrite their profiles continuously as a part of identity performance. Also, on the lifestyle blogs, new posts are often part of identity performance, and images increasingly become an important part of that. Susanna commented that sometimes if she does not have any images on her blog site, nobody writes comments. Therefore, bloggers need "photo publicity" in order to get attention. Another blogger at the school, 14-year-old Kaisa from another class, told me that she hoped other people would be interested in things she is doing and that the blog would serve as a virtual business card to possible employers:

> When we were in the flea market selling bags that I made, I wrote my blog address to the bags. There was somebody who bought my bag and she has a company that renovates old clothes. My parents advertised my blog address and that woman promised to go and have a look at my blog. (Interview with Kaisa, 14, female)

Kaisa used her blog to share photos of herself, her journeys, parties, and everyday life. She presents her clothes and writes short posts about her life in Finnish and in English. She describes her blog as her "very own corner," where she saves elements from her life. She is also an active blogger; in 2010 she had 92 blog entries.

Kaisa's description of a blog as a "very own corner" indicates that a blog is individual and made by one person. But blogging can also be a way to participate in the online network of people and other bloggers. Both Kaisa and Susanna mentioned famous bloggers with hundreds of followers.

Bloggers also share their knowledge of web design and learn from peers. Susanna said that she had designed about seven blogs. She had learned a little bit of HTML coding in order to design blogs. "I have tested things myself and asked something from my friends," she answered when I asked how she had learned to code websites. Kaisa had learned HTML coding as well: "I found everything I need from the Internet. My old blog went totally down and I had to start from the beginning. But now I have learned that, I do not do the same mistakes again." Kaisa had modified a code on her old blog and something went wrong so that she could not fix the code again. So she started again and learned how to do it. One of the other girls mentioned that she had designed a web page at school and learned some basics of HTML coding, but Susanna and Kaisa did not remember any teaching period of HTML coding at school.

I observed some joint classes of history and ICT of seventh-grade students who designed web pages in the computer class. The school had new web design software installed on the school computers, but students had not used it before. Three of the 24 students said that they had designed web pages before, but by using HTML coding. One of the boys said that he had created a discussion forum for a game site. Web design is not a priority in the school curriculum, but ICT teachers can at least give a hint of the practice. Students who want to learn more have to develop their knowledge and skills elsewhere, and in most cases, this comes from the Internet and peers.

· 5 ·

CLASSROOM STRATEGIES

Reading and Writing the Media

In this chapter, I focus on curriculum-based media production in official school spaces, where learning about the media was situated. Buckingham (2003b, p. 70) referred to six basic teaching techniques that usually suit school curriculum and official school space in media education context. These are textual analysis, contextual analysis, case studies, translations, simulations, and production. All of these techniques take place in Finnish schools. I first describe some practices of simulation, production, and textual analysis in the mother tongue class, especially in the context of analyzing and making advertisements, which is often used in media education. As previously mentioned, one of the main goals of media education is to educate critical consumers. This was also the main goal of the advertisement project for the eighth-grade mother tongue class.

Textual analysis is one of the most frequently used techniques of media teaching. As Buckingham (2003b) wrote, textual analysis means "making the familiar strange" (p. 71). In the mother tongue class, the students' assignment was to make some kind of close reading of the meanings of the images, film, and other media and audiovisual texts as well. The goal was to understand how a text has been put together and how it constructs meaning. Buckingham pointed out that textual analysis usually has three steps. The analysis begins with a description: "Students are asked to identify and list everything they can see and hear in the text." The focus is partly on the use of settings, color, body image—what is shown—and partly on how it is shown: the use of shot angles, fields of size, composition, lighting, and so forth. The second step is to consider the meaning of the text, the third is to make judgments about the text, and all of this is done in a systematic way (pp. 71–72).

These steps were not clearly separable in the advertisement project and were not done in a systematic way. Before the first lectures, students were asked to view TV or magazine advertisements at home. The assignment was presented in a schoolbook in the following way:

Choose a youth magazine and explore ads.

- Give all information about the magazine and present it.
- Find five ads that you suppose to be targeted to you.
- Explore the ads you have chosen and answer the questions: what kind of look should you have according to the advertisement? In what kind of environment should you be living? What kinds of products should you want to have?
- Make a description of young people to whom ads are targeted.
- Make a short report about TV ads.
- Choose one sport, children, or food program that includes advertisements.
- Look at the advertisements and make a list of products that have been advertised.
- Categorize the products.
- Explain to what kind of people ads are targeted: Who are they? What are they thought to value and want? In what kind of environment are they living? Who belongs to their life? What do they do? (Mikkola, Luukka, & Ahonen, 2006, p. 67)

The assignment did not have the clear steps described here and there was no specific approach of media teaching, teaching about the media, and no single question about how the visual language is used and meanings are made. One reason for this lack is that culture and media studies do not have as strong a role at schools in Finland as they do in the United Kingdom as a media educational approach. But still, the teacher and students discussed the language of advertisements, and some steps of critical media analysis were made in their school textbook:

In this way you read an advertisement critically:

- Look at what is advertised.
- Explore, to whom an ad is targeted.
- Think what things are thought to be important for the target group according to the advertisement.

- Explore what means are used in the advertisement:
  - Lingual means
  - Audio-visual means
  - Content selections
- Explore what the advertisement does not tell about the product.
- Make your own opinion of the product. (Mikkola et al., 2006, p. 69)

As visual means, the book presents image sizes, colors and shadows, and shot angles. But the emphasis in the assignment was the target group, audience, and how advertisements speak to different audiences.

Production, for example making an advertisement, is a practical hands-on use of media technology (Buckingham, 2003b, p. 82). Students "write" the media text with different techniques, in this case by video camera (video advertisement) or paper, photo cameras and pencils (magazine advertisement). As Buckingham (2003b, p. 82) emphasized, "reading" and "writing" the media should be inextricably connected. Media and digital literacy does not only mean "reading" and interpreting media texts but also producing them in different contexts.

Because *literacy* refers (in Finnish) so clearly to written texts, we have some difficulty using the term *media literacy*. As we have seen, one problem is a cultural dichotomy between visual-digital media and printed media. In common discourse, visual and digital media belongs to the world of leisure, home, and children, whereas written text belongs to the world of school, culture, and education. Written text, at least from an educational perspective, is often prior to visual and multimodal text. In this context, literacy is associated with high civilization in which the written word has an essential meaning and is a basic tool in education.

This dichotomy between different literacies reflects a traditional dichotomy between language and picture or word and image. Image is often seen as a "natural" sign, which does not need to be read. For example, Paul Messaris (1993) argued that understanding the kinds of visual conventions used in most mass-mediated movies or TV programs does not require any kind of visual or media literacy, because pictorial signification "is built on our everyday, real-world skills of physical and social perception" (p. 40). From this perspective, the main point is that visual modes of communication are misidentified as forms of language and literacy. Instead, for Messaris (1993), a word is a conventional sign and has a cultural code system. But there has always been a very complex connection between these different signs, words, and images.

In the Finnish school system, "text" is considered in a broad sense. In a teacher's book, a text is defined in the following way:

> Text in our times may consist of written and spoken words, images, image texts, moving pictures, sounds and sometimes even happenings. A text is more often multimodal; it uses different means and resources in order to communicate a message. The word text does not mean anymore a pure written product, like often in everyday language.

> Understood in a broad sense it can be what so ever meaningful totality. (Mikkola et al., 2006, p. 12)

This definition of the text is clear, but it is not accepted among all teachers and researchers. In practice, it is used in school teaching, but the mother tongue teachers in the Maple School did not use the term *media literacy* or even *media education*. Like the textbook suggested, the question was about text impacts, how different (media) texts impact, in which way, and to whom.

In the advertisement project, students took or simulated the position of advertiser. As a technique of media education or media teaching, simulation is a form of role-play where students are put in the position of media producers. Buckingham (2003b, p. 79) argued that it is particularly useful for addressing questions about production roles, processes of media industries, and technological, financial, and institutional constraints of productions. Simulation offers opportunities to understand the nature of representations and audiences when students, for example, produce their media productions from a particular perspective and for a particular audience.

Making an Advertisement in Mother Tongue Class

I was observing several different teaching periods of media literacy during the academic year. One of the most interesting was the aforementioned advertisement project in a mother tongue class with eighth-grade students. The assignment for the students was to design an advertisement campaign for a product called *Jyty*. Jyty does not mean anything in Finnish, but it has some onomatopoetic connotations to something strong and powerful. The project required students to design the product that was targeted to young people, produce an advertising campaign, and create an advertisement either in print or video format. The goal of the project was for students to understand persuasive techniques, impact, language, and rhetorics of ads and concepts like target audience. The assignment was a combination of simulation, production, and analysis of the text. Students simulated an advertising agency but not any specific professions, designed a campaign, and produced an advertisement.

The teacher gave written instructions for students:

> **Advertisement project, 8th grade**
>
> You are working at the advertisement agency (find a name for your agency!). You got a job to plan a campaign for a new product targeted to young people and called Jyty. Jyty is produced in Finland. As experts, teachers and students from the Department of Design at the Aalto University are used, so it is a completely Finnish product.

For an advertisement campaign you should write:

- Description of the product: a possible information label and technical information
- A plan for marketing the product in Finland (how many ads, channels, timetable, point of time, duration, etc.)
- What will be told about the product
- A draft of a magazine ad
- A draft of a television ad

The starting point for the project was the theme of impact modes in advertising messages. In fact, advertising messages were included in the broader study of propaganda, persuasion, and impacts of language and media. This broader context included that students studied argumentation, writing letters to editors, media knowledge, debates, literature, and literacy skills. Media analysis and especially a formal study of advertising were included in the eighth-grade mother tongue textbook. Before the project, students read from the textbook examples of how advertisements and images make meaning, what kinds of means are used in ads, and how advertisements address audiences. The textbook also included some writing assignments.

In the beginning of the project, the teacher also taught impact modes in advertising; She talked about "language and rhetorics of advertisements," how ads have an impact within language on audiences. She discussed with examples how ads appeal to audiences' emotions, how authorities and celebrities like athletes, doctors, and scientists are used in promoting the quality of the products, how foreign and academic language and slogans are used in ads, and so forth. The teacher asked the students questions and they tried to answer as they usually did in school learning settings. This is the classical classroom-teaching, where the teacher is a director and draws the class in his or her direction.

The time was limited and there was no opportunity to go into deeper dialogue, even though some questions caused lots of comments. The teacher explained how some trends are used in ads and gave an example of car advertisements, where the trend is to be environmentally friendly: "The more environmentally friendly the product is, the more definitely this argument is used in an advertisement," she taught. One of the students asked the following:

Kari: Why must everything be "green"?

Teacher: It is because the world will be saved.

Jouni: You know that the whole thing is verified so that only a part of the researchers in the U.S. have that kind of opinion and all other researchers all over the world have to agree with that.

Teacher: What do you mean, what do they have to agree?

Jouni: That climate change is coming.

Kari: My opinion is that carbon dioxide is not harmful for the atmosphere. I have a right to that opinion and I am allowed to drive a car as much as I want to.

Jouni: It is harmful for the atmosphere, but if we are against climate change we must kill all dogs...and cows and save all four-wheel drive cars.

Teacher: Yes, it is true that this is a very complicated and diverse thing, but if we now stay here in this topic of advertisements. Ads argue with facts as well.

Vesa: These climate friendly cars are so expensive. People would buy them gladly if they were much cheaper.

Teacher: Yes but cars have emission criterions, how much carbon dioxide is allowed. Nowadays the tax of a car depends on how much emissions there are. Cleaner cars have smaller taxes.

Kari: Teacher?

Teacher: Yes Kari, this is the last one, then we shall move on.

Kari: Think, that if there really will be a climate change, people would have found a solution after all...there would be no cars anymore if this is what is really wanted. But cars for example are so big business.

Teacher: Yes, that's true.

Jouni: Politicians make money with climate change.

Teacher: Yes, that's true as well. You should take an environmental point of view to your advertisement.

This short episode shows that discussion about advertisements gives some room to broader ethical and political discussions, where the theme is close to students' everyday life experiences and knowledge. Talking about ads is based on real phenomena and not abstract school texts. This is important from the point of view of literacies. Roger Säljö (2001), for example, has argued from the perspective of sociocultural learning theory that learning settings at school are usually out of students' lifeworld and belong to the sphere of abstract academic thinking. School literacies are focused on texts just like other institutional literacies. Säljö gave the example of Newman's study from 1982. In the example, Bert and Ernie, two characters from *Sesame Street*, have a discussion about sharing a banana (Säljö, 2001, p. 198). Ernie has a banana and starts to eat it but Bert wants to have a part of it:

Ernie: Part for you and part for me. Yes indeed, Bert. I'm going to divide this banana right now, so hold on a second.

Bert: That's nice.

But Bert sees that Ernie is eating the whole banana.

Bert: Ernie, you are eating all that banana, you ate it all.

Ernie: No Bert, we are sharing it.

Bert: Sharing it?

Ernie: Sure. I took the inside part and here is the outside part for you.

(Sesame Street, 1972)

In Newman's study, children of different ages were asked if Ernie lied because he only gave the peel to Bert. Younger children admitted that Ernie lied but older schoolchildren said that he was both lying and telling the truth. Ernie promised to give part of the banana and the peel is part of it. Säljö (2001, p. 199) pointed out that this is a common way to deal with arguments in school learning. The same kinds of textual practices are used in bureaucracy and juridical contexts. Institutional literacy practices divide the logical and communicative aspects of arguments. This helps to speak about things that are not present and perceived immediately and are not part of communicative contexts. This is the textual realm of school subjects and textbooks. From the perspective of sociocultural learning theory, communicative contexts are as important as logical ones and it is possible to take them into account in learning. Speaking about advertisements, for example, is a communicative context. Learning is tied to embodied experiences in the social and material world and not only to abstract concepts (Gee, 2010). The teacher's role in situated learning (Gee, 2010; Lave & Wenger, 1991) is to be more like a guide or coach rather than a director and give more room for students' own ideas and their emotional investments in the media.

Situated learning had room especially in group work during the advertisement project. The group work started when the mother tongue teacher gave the assignment for groups. She emphasized that it would be good if students considered environmental topics, such as recycling and energy saving, in their advertising campaign. The initial ideas within the groups were quite innovative technological devices: a dishwasher that cleans itself, a remote control than can be used as a phone, a massage bathrobe, a vibrating dish brush, and so forth. The teacher wondered how these are targeted to young people. She commented to me later that these kinds of creative assignments usually begin with quite a mess. Students argue with each other and noise rises when they try to find out what they will do, how and who will do what. She told me "a comprehensive school is a place where students have the right to do things in practice, and not only read from the textbook something about advertisements." She added that this kind of learning by doing is no longer common, that "books are heavier," and that there are more and more themes to teach.

One group decided to make a television advertisement of the Jyty sport drink. Simo tried to find a slogan: "Jyty gives you brains."

> *Tarja*: I think that we can make [a video] if we can do texts by Movie Maker in time.
>
> *Teacher*: Yes, everybody can use it [Movie Maker] at home.
>
> *Hannu*: How many hours do we have to do the video?
>
> Teacher: Three hours.
>
> *Hannu*: Three hours, okay.
>
> *Teacher*: Then your video is ready. Remember that if you do a video you have to have a product. You have to have a bottle from where you drink and a label, where is that Jyty text and that has to be ready next week.

Students began to consider which one is better, a bottle or a can. Simo called out that he could design a label with his computer at home.

This vignette shows that students began to think what could be done at home and with their own computers. They did not have much time in the official school space and they had to do some work off-school and off-timetable. Most of the groups later said that they did not have enough time and that is why they had to make some compromises. When students are making their own projects the time flies and the space-time structure of the school is not enough. Teachers know this very well but they are not able to change the structure:

> "Now it is time to start to clean places, and on Wednesday we shall continue," the teacher said. The students begin to gather paperboards, pens, cameras, cell phones and other stuff. One group is still in the school's gym where they are shooting their video. The teacher tells me that they have lost their sense of time in the gym. One hour is really a short time. I ask what happens if they stay longer at the school? The teacher answered:
>
> Students complain about that and there will be hooky playing. It depends what you are used to doing. If you do not know anything better, there will be no problem. These 45 minute classes. You try to find some extra material and take it into sight and immediately you have to start to take it apart. It really is sometimes frustrating. In the upper secondary school we had 75 minutes and it was much better. (From my notebook)

Time and space are always a problem, as discussed earlier. The school architecture and teaching schedule are not suitable for media production. This is one reason why teachers prefer books and writing assignments—and students make their productions off-school and also publish their schoolwork on the Internet. Within traditional school settings, it is not common for students to be so proud of their schoolwork that they want to present it in public. But media productions are an exception; part of the nature of media is that they are published and circulated. When the aforementioned Jyty sport drink advertisement was ready, one of the group members published the video on his YouTube channel. The group even presented their advertisement in the mother tongue class by using YouTube. The teacher made some comments about the public publishing of schoolwork, especially from the point of view of copyrights, noting that everybody who had a part in making the video should have given their permission for the publishing. The video did not include any music and thus music copyright was not a problem. The group members commented that they had an intention to make some music themselves, but they did not have any opportunities to do that at home.

Group members also presented their imaginary campaign plan, when and how often they intended to show their advertisements on different television channels and some details about the Jyty product itself. Some groups had details from their production, for example storyboards and special informative labels for their product. A group that designed a magazine advertisement campaign for Jyty breakfast cereal presented a rap for the product in the class. Others had made composition and music for their video with their own lyrics as well. Students understood that music is an important part of an advertisement, even though the teacher did not mention it in the beginning of the project.

When groups presented their campaigns and advertisements they used language that appeals to health, humor, technological excellence, nationalism (made in Finland), environmental friendliness, and so forth. These were also highlighted by the teacher at the beginning of the project. Students showed that they are familiar with the basic language of advertisements in general, but they had some problems conceptualizing what impact modes they used in their ads.

At the end of the project, every group had a feedback meeting with the teacher where they discussed the group work, problems, students' participation in the project, their products, and language and rhetorics that were used. A group that made a video advertisement of a weight loss cake understood the idea of weight loss products well. The teacher asked group members how the product and the ad were targeted to young people. Girls in the group answered that young people have a lot of pressure to look good all the time, but they admitted that they did not remember to take young people into account in the advertisement. In the video, they played the roles of middle-aged women more. Actually, weight loss products are usually targeted to people in their middle ages and it is more or less illegal and at least ethically problematic to target these kinds of products to young people, because of the eating disorder issue that many young people face. Thus, girls in the group used traditional rhetorics that they have seen on ads before. When the teacher asked how the language of advertisements worked in their ad, the girls remained silent for a long time.

> *Teacher*: Did you use effective impact modes of advertisement language? We discussed about these modes in the beginning of the project.
>
> *Mari*: I do not remember those.
>
> *Teacher*: You wrote those down in your notebooks.
>
> *Mari*: What were they?
>
> *Teacher*: There were eight or something; humor, appealing to emotions, authority...
>
> *Satu*: We used humor...
>
> *Teacher*: Humor, yes you are right...and you used before-after structure.

All of the groups found it difficult to articulate and conceptualize the language they used. One problem with hands-on productions is that students easily become lost in their experience and practice. Therefore, the teacher also wanted to go back to conceptualize means that students had used in their ads. Students are already knowledgeable and familiar with advertisements, but the assignment was supposed to "make the familiar strange" by deepening students' conceptual understanding.

Burn and Durran (2007, p. 153) made a reference to Mimi Ito's idea that media literacies are partly learned, like spoken language in the peer-to-peer social ecologies and in cultural practices. They argued that students' familiarity with media culture and their own engagement with media production outside of school is something that teachers should meet halfway. Students' media tastes and skills will meet the school culture of media literacy and this "meeting" is important to recognize and supervise, for example, in curriculum-based media productions. Burn and Durran (p. 152) suggested that this kind of "dialogue between cultures" can progress media literacy in cultural development. This may be one reason why advertisements have such an important role in media teaching at school: they are part of everybody's experience and are easy to adopt to the school and to the context of critical analysis. Additionally, advertisements are a common area for teachers and students, and media education has long paid attention to the effects of advertising and consumer culture (Buckingham, 2000, p. 145).

Critical Media Literacy at School

Media culture is continuously changing and advertising is more complicated in the age of the Internet than in the age of mass media. Joseph Turow (2011) reminded us that the consumer certainly is not a king in the new media environment, like Nicholas Negroponte (1996) and others had suggested. For Negroponte, the digital media environment would offer power to the people, allowing them to focus their interests on what they value. For example, by using the term *participatory culture,* Jenkins, Clinton, Purushotma, Robison, and Weigel (2006) have also focused on active audiences and their growing influence on the digital environment as facilitators of civic engagement. Turow argued (2011) that even though this is partly true, "the centrality of corporate power is a direct reality at the very heart of the digital age" (p. 7). Marketers and ad agencies profile and value their audiences and use audience targeting and database marketing for their purposes everywhere on the Internet and social networking sites. New spaces where people may participate and create content are at the same time "constrained and channeled by industry logics shaped by new media-buying mindsets that speak directly to whether, when, and how items are presented to particular audiences for attention" (p. 18).

Schoolbooks do not tell anything about the new logics of marketing. The focus of books and teaching is on mass media like TV and magazine advertisements, newspapers and movies, and the logic of online marketing seems to be largely unknown in the context of school curriculum.

In Grade 8, the students also wrote an essay about advertisements. The teacher gave students some headers for the essay, but once again the context was mass media. According to the teacher, the most popular headline for essays was "TV ads—unnecessary complement or necessary advertising?"

Students tried to analyze TV ads and their meaning for people and society. Some students seemed to still think that advertisements are some kinds of windows to the world and they just present the product as it is. They did not understand the nature of representation and how advertisements offer reading positions for the audiences. Others had a stronger sense of critical reading and media literacy. They wrote how advertisements use humor, slogans, and exaggeration, manipulate emotions, and play with gender roles. Female students especially were interested in gender differences and picture manipulation that they saw in hairspray, shampoo, and makeup advertisements:

> It seems that almost in all advertisements women are presenting products. Men never advertise hairspray or shampoo. They only admire the impact of new hairspray or shampoo.... The casting in advertisements where only a woman is present makes me wonder do men and women not get along with the same products? Are men not able to use for example the same hairspray? (From Grade 8 essays, female)

> A make-up advertisement is possible to identify in two seconds when beautiful women step onto the screen in a very well-lit make-up studio or in a city landscape. In the advertisements women are young, skinny and they have perfect skin already before the make-up. It is made easy by modern image processing. (From Grade 8 essays, female)

In the aforementioned advertisement project, one group changed gender roles in the magazine shampoo advertisement group members created. Girls made an ad and the only boy in the group got a role of the model. This gender blender in the advertisement was interesting, although students used traditional and stereotypical means in their advertisement (before-after structure). When they presented their advertisement to the class, they could not tell what means they had used and did not notice the changed roles in their ad.

In the mother tongue class, the emphasis was on the language and rhetorics of advertisements. There were differences between students on how they understood this language. TV and magazine advertisements are one of the easiest ways to start exploring media texts and enhance media literacy. Students are familiar with ads as part of their everyday experience, but the theoretical framework and concepts are difficult to use in practice.

Previous research has noted that while young people may have more knowledge about advertising, they may not necessarily employ critical thinking skills in their understanding of advertising (Hobbs, 2004). But these findings are controversial after all. In her study of media literacy in Grade 11 English language arts in the United States, Renee Hobbs (2004) found that students who received media literacy instructions gained increased knowledge of the production processes used in advertising and their ability to critically analyze advertising grew. They, for example, demonstrated the ability to support their arguments with evidence concerning target audience of the advertisements.

Students in the mother tongue class at Maple School also indicated abilities in identifying the target audience, especially on the grounds of magazine target groups. One of the eighth-grade students wrote the following in her essay:

> Different magazines have advertisements of products that are targeted to the special group of readers. For example, in women's model magazines are ads of branded clothes, perfumes and cosmetics. In a [local newspaper] are ads of cars and electronics, because men are reading it as well. Youth magazines have ads of music, electronics, food, clothes, etc. And this applies also to other media than magazines. (Part of essay of eighth-grade girl)

In discussions with the teacher, students pointed out what kinds of items are targeted to young people in various media. Some students clearly thought about advertising and used critical concepts when they analyzed ads and discussed the media. But my sample, of course, is not statistically valid and there was no control group. Therefore, it is not possible to tell if media education and the advertising project in the mother tongue class had any specific influence on students' media knowledge or skepticism toward advertisements.

In their essays about advertisements, students also wrote about their consumption habits, how they use branded clothes because of better quality, and how advertisements have an impact on their everyday life. Media education aims at forming young people as critical consumers, but they find it difficult to change the role of bare consumer. Ahonen and Virta (1999) wrote the following: "There is a *contradiction between ideals and practice* of media education as there is in civics in general" (p. 248). Students have good knowledge and they know things in principle, but they do not adopt their knowledge to their everyday life.

Critical and conceptual analysis of the means of advertisements or any other media text is not easy to make at school. Students do not easily adapt academic standards and concepts without a clear bond to their lifeworld. Teachers, for example, use a lot of material from newspapers because it is easy to bring to the class. But young people prefer other media and may have sarcastic attitudes toward newspaper activities in the classroom.

One of the mother tongue teachers at the school had a short project for ninth-grade students entitled "Critical media study." The teacher discussed how journalists make choices when they write news and how the objective and neutral perspective is difficult: reporters take the viewpoint either intentionally or unintentionally. His example was diction: it makes a difference if a journalist chooses the word "liberation" and not "attack" in the case of an activist attack on a fur farm. Students did not have much critical discussion, but they did complete an assignment given by the teacher. The assignment was to analyze propaganda, write a short news story for different newspapers, and explore how an image in a newspaper changes the message of a news story:

> Choose a news story from a newspaper and either modify an image or make a new image that changes the meaning of the news and especially the headline of the news story. Remember to write a new caption! (An assignment by the teacher)

Students got some newspapers from the teacher's desk in order to read news and cut some pictures from the newspaper. The teacher said that the school received about 20 copies of *Helsingin Sanomat,* the leading newspaper in Finland, every day, but he did not know why: "Most of them go straight to the paper recycling."

Finnish people are quite active newspaper readers and PISA success has showed that young people who read newspapers regularly do better in PISA exams (Sanomalehtien liitto, 2012). Still, there are students who do not read newspapers at all or follow news. For example, I interviewed a 16-year-old student who said she does not follow any news: "I do not [laughing] read newspapers. Usually I hear from other people if there is something going on, what has been said in the news."

In the mother tongue class, the newspaper "cut-and-paste" assignment led some students to sing their frustration with their own tune:

> cut and paste and manufacture
>
> cut and paste and manufacture

After a discussion about last night's ice-hockey game, some of the boys shouted, "I can't stand this anymore!" The boys did the assignments mechanically and were not involved in critical thinking at all. They simply cut out a picture of a bald man from the newspaper and wrote the text, "it's great to be an adult man—the bald is shining." Others shouted that they could not find any news from the newspaper. Boys giggled together: "This has absolutely no sense."

In an interview, Riku, age 14, said that in another class he was tired of just viewing advertisements hanging on the wall and making analyses: "It would be much more fun to do something and not just go through ads." When I asked if he found analyzing advertisements interesting at all he answered unequivocally, "no." Sometimes teachers had to force students to do some analysis. One example is from the movie course where students were supposed to do an analysis of the Finnish film, *Beauty and the Bastard* (*Tyttö sinä olet tähti,* 2005). In the beginning of the lesson, the teacher explained what film narrative means and how it works and gave students some questions to answer. Instead of doing this, students preferred to continue their own video-making process and asked if it would be possible to edit their own videos at the same time. After some negotiations, the teacher decided that students had to concentrate completely on the analysis.

Official school space and formal school settings do not support participatory hands-on education pedagogies, which is why teachers choose teacher-directed modes of education with academic text analyzing models. My study indicates that students have the best learning practices by using some kind of combination of analysis and production, like in the advertisement project. Hands-on projects are not influenced so much by representational theories of media education, which focus on the constructed character of media production and reception and ideological dimensions of the media (Asthana, 2011, p. 52). Rather, media productions bring out a nonrepresentational theory of media education, which focuses on complex issues of embodied practices, affects, and narrative identity (p. 52). This analytical level is a supplement to the previous one and is seen, for example, in how students perform in their videos and media productions and how bodies, for example, become a focus of meaning making and can articulate specific forms of resistance (p. 62). In the school, it may be seen in the production of unofficial school spaces and tactics, where students create a new social order and occupy the school space.

Soap Opera

Critical media literacy was in focus in other mother tongue classes as well. In Grade 7, the theme was soap opera. The mother tongue teacher used different pedagogies to explore soap operas with the students. In the traditional way, the class started with the discussion and textbook reading. The seventh-grade textbook in mother tongue included a learning area entitled, "Media and literature knowledge" (Mikkola, Luukka, & Ahonen, 2005, p. 195). Part of the area was soap operas as storytelling. The textbook included a rather long article about the history, structure, and characteristics of soap opera and even some findings from Finnish researcher Tuija Virta (1994). The academic reference is rather old in a textbook that was published in 2005.

In the beginning of the class, the teacher drew a mind map from the textbook on the blackboard. The mind map included some typical characteristics of soap operas, and the students added more. They discussed product placement, milieu, stereotypes, and cliffhangers of the soap opera. Students easily found examples of these from ongoing Finnish soap operas. The teacher gave some older examples from *Dynasty* and *Dallas*, which were unknown series for young students at that time. They analyzed show time, characters, and other dimensions of soaps. Students summarized soap opera in the following way:

> *Student*: They [character of the soaps] are hanging around with each other and falling in love with the neighbor living in the same block of flats, then leave that person and become pregnant and start a new relationship.
>
> *Another student*: And she is killed meanwhile.
>
> *Others*: And she comes alive again and notices that she had a relationship with her sister and brother. And has a heart attack when she hears that they are her sister and brother.
>
> *Teacher*: Okay, a kind of causation, chain reaction.

The teacher also asked some questions about the typical plot of soap operas and why soap operas are filmed only in some particular places and indoors. All of this was explained in the book article and the students had evidently read the text:

> "There's no clear time structure, old and new get mixed."
>
> "There's no clear climax."

To the question of filming indoors, one of the students referred straight to the textbook: "There was in the text that when you watch the program, it is like being at home, a familiar feeling."

Soap operas are popular TV programs in Finland, even among children and young people. The daily Finnish soap opera *Secret Lives* (*Salatut elämät*) has been on the air since 1999 and every Finnish person knows at least something about it. It has been under public debate several times. For example, after the season finale in 2009, when a pregnant woman was buried alive, the Minister of Communications questioned the show's suitability for young children. Students also remembered an episode where a restaurant exploded and nobody knew which of the main characters was killed. The show has dealt with themes like abortion, drug abuse, incest, alcoholism, eating disorders, and so forth, that have been under public debate. However, or therefore, the series has been popular with schoolchildren, from eight-year-olds on up.

Secret Lives was part of students' everyday life and popular culture. Therefore, it offered a basis for questions of personal opinions, taste, and affections. From the questions included in representational media education, the teacher moved ahead to the nonrepresentational questions and asked from the basis of research findings in the textbook why the students are hooked on soap operas. The majority of the students gave plot and characters as reasons, but some said that watching soap operas is a kind of everyday routine. The teacher himself confessed that he was hooked on *Secret Lives* for about 1 1/2 years. The teacher was one of the youngest teachers at the school and had a strong relationship with popular and media culture. He was able to share common experiences with students. Sometimes a generation gap obstructs a dialogical discussion with teachers and students, especially when it comes to life experiences and media culture. But in this case, the teacher (who was also known for his heavy rock interests) easily found examples close to students' everyday life.

At the end of the class, the teacher gave group work from the textbook. The assignment was for groups to generate an idea of their own soap opera and present it to others in the class. The teacher suggested that students should dramatize one scene from their soap opera and play it in the class. The next class was reserved for students' presentations and dramatized acts. Drama as a pedagogical means opened some kind of combination of official and unofficial school space.

Drama education does not have any official position as a school subject in the Finnish core curriculum, but it has been an important part of the school system in spite of that (Toivanen, 2012). Drama is placed within the school subject mother tongue and literature as a subarea of interaction skills. But it also has some meaning within media education as well, as in this case. Students had to think of and explicate some typical characters of media representations and apply these to their own presentation. The teacher said that teaching offers some means for analyzing the text, and that students know much, but "usually they do not know how much they really know and figure out."

Students were extremely motivated to conceive and dramatize their soaps. They designed some spinoffs to their soaps, web pages (only on paper), music themes, and plots. The teacher said that these kinds of assignments usually work with this class but not with every class. The students were used to the use of drama as a pedagogical means and were not afraid to act.

The plots were as complicated as in real soaps and the relationships were confusing and humoristic. In the soap entitled, *Secret Kebabs*, students explicated the series in the following way:

> The series tells about young teenagers who are in the middle of complicated relationships. We are able to see humoristic and dramatic moments. The series follows also their parents' work at the restaurant. The story begins when an exchange-student Alex comes to Finland to the family of Chris and emotions begin to catch fire. How do the parents and Lisbeth react to the relationship of Alex and Chris and who is the jealous intriguer? How their relationship and admirations affect the school life? What happens to the restaurant? Who is mysterious Aso? (From students' text)

Students used the "map of relationships" (Figure 5.1) in order to clarify who was with whom and who the people were. Characters were cut from a magazine that the teacher offered to the groups.

Figure 5.1: Relationship Map for *Secret Kebabs* Soap Opera

This soap opera seemed to be strongly influenced by the *Secret Lives* series; even the name was adapted from the series. This is once again an example of remixing culture that is an important part of young people's lives. This time the main characters of their own soap were gay. This alludes to some themes in the *Secret Lives* series but also to the liberal thinking of the group that included three 13- 14-year-old girls.

The group improvised in the front of the class a scene where Aso (son of the father, Halem) works in the restaurant of Halem and Melah (married couple and parents of Chris). The parents do not know that Aso is Halem's son. Halem scolds Aso because he was on Facebook during his working time. Finally, Aso confesses that he knows that he is Halem's son and all the emotions are unleashed.

Other groups had similar kinds of dramatized scenes. One group even composed a music theme, wrote notes to it, and recorded the music with a video camera. Girls in this group used their own video camera and brought it to the school for their presentation. Once again, the school's media production was done off-school.

In their dramatized acts, students used the classroom space differently than in normal class. They were embodied young people who presented their stories to their peers rather than to the teacher. Usually, schoolwork and tests are done for the teacher by an individual student. Now the school space was opened to a more participatory context where an individual learner was pushed aside and the mastery was not centered on the teacher. Students had a feeling that they belonged together and were doing something together. They laughed together and commented freely on each other's presentations. Each presentation was followed by spontaneous applause.

These kinds of dramatized acts are "participatory theatre" or "classroom drama," where the border between the performers and audience is obliterated (Toivanen, 2012, p. 229). Every student is in turn an actor and a spectator. For Jacques Rancière (2009), this kind of shifting of actor and spectator is an emancipative act: "That is what the word 'emancipation' means: the blurring of the boundary between those who act and those who look; between individuals and members of a collective body" (p. 19). For Rancière, the question is not so much about the theater but rather education, where students actively make meaning and participate in the teaching/learning processes:

> The spectator also acts, like pupil or scholar. She observes, selects, compares, interprets. She links what she sees to a host of other things that she has seen on other stages, in other kinds of place. She composes her own poem with the elements of the poem before her. She participates in the performance by refashioning it in her own way—by drawing back, for example, from the vital energy that is supposed to transmit in order to make it a pure image and associate this image with a story which she has read or dreamt, experienced or invented. (p. 13)

According to Rancière (2009), blurring the boundary between acting and looking opens a "community of narrators," an emancipated community where the traditional roles of the teacher and students have been changed and learning is seen more as active use of the voice and students' own life experiences. Media education, and here drama teaching, can be seen as a possibility to change these roles.

Movie Course

The movie course in Grade 8 offered possibilities for performative acts out of official school spaces. The movie course was an elective and students had different reasons for choosing the course. For example, Mira, age 14, wanted to choose Spanish but the course was already booked: "I had to choose something and with my good friend we just decided to take the movie course." She said that she likes movies but not so much the course: "It is a little bit boring because we do not do anything." She expected that they would watch some good films in the course, "the kind of films that everybody likes, something like *James Bond*."

The teacher confirmed that some students thought the idea of the movie course was to watch films "and to eat popcorn.... It is easy business when you do not need to do anything." Instead of *James Bond*, they watched other movies, for example, a Finnish film, *Beauty and the Bastard*, and a *Wallace and Gromit* animation. Mira commented that *Beauty and the Bastard* was boring, but she was not especially keen to do video movies either. In their group, she did not want to participate in the video-making process and rather read textbooks for school exams and used her mobile phone to listen to music. When the teacher asked if they could share some tasks in the group, Mira answered that she could not do anything, rather she just wanted to follow others. Gomi, another group member, tried to give some advice on how to use video-editing software, but Mira rejected it saying, "I'm not interested."

Some of the other teachers also reported to me that media production was sometimes a problem for students: "Usually the most difficult thing is production," said the ICT teacher. "It is no matter how fun it would be, for example cartoon; they don't get excited because they don't find a story." It is sometimes difficult to motivate them.

The teacher explained that in some years, the movie course was divided into two parts: some students made videos and others just watched movies. But this year they had a good class and the majority of students really wanted to make something:

> *Teacher*: I have to praise you [the class]. This is the first time for a long time that we have shot the videos and we can start to edit them. Usually it has

always discontinued, when we have started at spring term. It's worth starting in autumn. (From my notebook)

The teacher explained that in the fall the students are more energetic to start hands-on practices. The spring is more difficult, especially because of long winter darkness, that continues in Finland until the beginning of March.

Some students, like Gomi, were extremely motivated to make a movie. But she was the only active student in her three-person group with Mira and Jori. She selected the movie course because she had some plans for the movie business in the future. She thought that she could learn to use editing software better while gaining some information about film history. Unfortunately, she was not satisfied with the course because she had not learned anything new and had already seen the films that were screened during the course, "but it [the course] is going in a better direction," as she pointed out.

Gomi became interested in video making when her mother showed her film editor software on the PC when she was in the fourth grade. She also had some experience with video making from primary school, where they made an animation film. In secondary school, she made short animations for history class and social science as well. Teachers allowed her to use her visual strengths in order to make school assignments. As the visual arts teacher said, "She is a top guy and because she makes films at home too, she has her own vision and visual thinking skills."

Gomi spent a lot of time on her videos, and this was done out of school. Besides the vlog, she had another blog where she wrote about films she had seen. In 2010, she wrote 22 blog entries about different feature films. She also commented on the blog about her own video: "In the movie course we got our video ready, and it was presented. We got some applause D;."

Working alone in the group was not a problem for Gomi, but she did not like that others were always complaining about something. The teacher said that there are always students in the groups who do not do anything. She had an idea that therefore there should be more other assignments like film analyzing:

> Actually, I have a feeling that there were lots of things that we didn't have time to do. After all, the major part of the course was practical doing, and because it was groupwork, there are always students that do nothing. The only textual part was film analyzing when I was off. If I would be a mother tongue teacher this course would be different, more analyzing and script making, different emphasis. (Interview with the visual arts teacher, female)

The teacher emphasized her role as a visual arts teacher who taught visuals rather than narration:

> I'm such a traditional visual arts teacher, I teach shot distances, shot angles, and if we think video, how to make transitions...but the basics are that students learn to perceive

> the environment and things. Really essential is that when you have some medium, you watch the world from a different point of view, whatever the medium is. (Interview with the visual arts teacher, female)

As a traditional visual arts teacher, she did not think that she was doing any media education in the class. During the school day and in every class, there are always so many things going on at the same time that it is impossible to distinguish what is media education and what is not or how to segregate different things at all. Another visual arts teacher said that he usually does not know exactly what he is teaching, whether it be philosophy or visual arts or something else. This indicates that media education and media literacy are deeply intertwined with everyday teaching without separate courses. Both the teachers emphasized critical aspects in their teaching, how to see the world and act in it, but the official school space did not give them room to do all the things they wanted to. Both teachers emphasized that there are too many students, 24 in the visual arts class. This kind of group size does not allow students with different skills to really concentrate on their work: "This [visual arts teaching] is far away from anything normal, somebody could call it chaos" (interview with the visual arts teacher, female).

The movie course included some training with photo cameras, as a practice of shot angles, image framing, and narrative. After a short training session, a group of students got video cameras and started to make plans for their short film, *iScream*, that has its starting point in Edvard Munch's (1863–1944) classic painting, *The Scream*. The teacher warned that the camera is not a toy:

> This is the school's camera. It is not allowed to break down. This is not a joke. Whoever is shooting has the responsibility for it. We will stop immediately if cameras break down. We do not have any extra cameras.

The teacher explained to me that whenever the students get the camera, they begin to play with it:

> When they get the camera in hand they begin to run in the corridors and stairs. It's kind of playing; it [the camera] is like a toy, He-Man or something else. I remember one situation, we have some plastic fish heads [masks] here, and when I proposed to one class that we could make a video film with the masks, make a script, so that there would be two people who would venture somewhere, and we had to think what they do and in what context. What happened was that two boys took the masks and began to fight and the other one got a scratch or something. That was the end of our video class. The funniest thing was that other students shot by their mobile phones when the two boys were fighting with fish masks on their head. I said that do not upload these to YouTube and asked them to delete from the camera what they filmed. (Interview with the mother tongue teacher, female)

The teacher pointed out that the students have some kind of inner thought that the meaning of the movies is fighting: "When they take the camera they have in their heads a model of action movies and copy it." In fact, children and young people usually start to imitate action movies at an early age. Therefore, some teachers are skeptical about the educational value of media productions. They do not see that students are doing more than imitating mainstream media, especially in its most violent forms.

Teachers also keep in mind the distinction between low and high culture that has been under long debate in the field of media education since the times of British literary critic F. R. Leavis and his student Denys Thompson, who represented the first systematic "curriculum" about the teaching of mass media in schools in 1933. Leavis and Thompson saw popular media as a corruptive force that crashes traditional and national values of great art and literature. Of course, teachers today do not follow in these footsteps, but the debate was alive until the 1980s (see, e.g., Buckingham, 2003b, p. 124). Buckingham (2003b) suggested that the concern was motivated by a fear of imitation that was seen to be an unthinking process. But as he argued, studies have shown that

> students' uses of popular media forms and genres frequently display a clear understanding of "media language," and a form of ironic distance that is at least potentially "critical." By making these dimensions of their work explicit, and through subsequently reflecting upon them, students can be encouraged to develop a more thoughtful approach to concepts such as representation, which are sometimes dealt with in rather mechanistic terms. (Buckingham, 2003b, p. 124)

Reflecting upon these aspects of students' work in the Maple School was explicit in some cases where teachers discussed with the students their videos and other media representations. Still, some teachers expected students to be more "artistic" in their video productions. One way to teach media is to mix popular culture with classical art. For example, in the visual arts class, students mixed Hugo Simberg's (famous Finnish painter, 1873–1917) art with imagery from popular culture. In the mother tongue class, students adapted the imagery of Kalevala[1] to modern advertisements.

Violent images and some kind of "trash" media can also be seen as an expression of tactics and alternative spaces outside the school. Some of the boys in the movie course filmed the aforementioned movie *iScream* with a narrative where two boys find a gun from the bin, run away, and get shot by a third boy. The whole story is fragmented and also repeated backward with music adapted from the *Benny Hill* series from the 1990s. As such, the movie is not an imitation but a satire and an ironic comment with the title *iScream*, where the boys create a parallel and disordered space besides a more structured school space. In their film, even the path of time changes, breaking the more future-oriented school time and hierarchy.

---

1 The Kalevala is the national Epic of Finland, compiled by Finnish physician, philologist, and collector of traditional Finnish oral poetry, Elias Lönnrot, in 19th-century Finland and Russian Karelia.

The output of the "Scream" assignment in the movie course was four short video films where only one included a battle with guns. Others were more peaceful stories. Gomi's group made a video where a man is building a huge track with domino bricks that breaks down. In the last image of the video, we see a scream. The students presented the simple story visually with different shot angles and music that represented how time goes on. In another story of another group, we even saw the fish masks. This was a rather surrealistic story of a secret laboratory of fish heads who make some kind of chemical in order to transform human beings. All the videos were constructed using basic visual storytelling with different field sizes, shot angles, slow motions, and speed-ups. The short 2-minute movie format gave possibilities to experiment with visual storytelling.

When students made their video films, they opened new spaces to the school and went to shoot in the church garden next to the school, to school corridors, and the cellar. They were motivated to shoot and edit their films, but they were frustrated when they had other assignments to do. In February 2010, some students suggested doing a feature film, but the teacher had other practices in mind. She divided the class into three groups and gave students newspapers. The assignment was to search for newspaper advertisements and news and cut them out.

Riku: Why are we doing this, what's the sense?

Teacher: This is a dry run.

Janne: What will we do with these [images]?

Teacher: First we collect images, then we explore them.

When somebody asked how many ads and news they had to collect, Riku called out, "I do not know. I just cut them. Who cares?" He showed his deep frustration by picking up his headphones and starting to listen to music. The teacher asked him to stop the music but Riku did not react at all. "Stop the music!" the teacher demanded, and Riku finally took his headphones off his head. Others started to read image texts and guess without seeing the image whether it was a piece of news or an advertisement. The goal of the assignment was to explore differences between news and ads and later to shoot a short video advertisement.

Once again, newspapers were the main material for the practice and, like in the mother tongue class, some students protested strongly. They seemed to find working with newspapers, cutting and collecting images, to be a childish activity. In the same way, a film analysis caused strong protests:

"We feel bored because we have discussed so much."

"Can we watch a movie?"

"Some horror movie, please?"

This time the assignments were given by a substitute teacher. When she started to ask questions, students stayed quiet. The teacher tried to ask what was the key line of the film they watched last time (*Beauty and the Bastard*, 2005, Finland) but the students answered, "What do you mean by key line?" and "Shall we watch a movie? We do the analysis at the same time, really." The teacher had promised that they would watch a *Wallace and Gromit* movie *The Wrong Trousers* after they had discussed the analysis. She then repeated that *Wallace and Gromit* would be a "prize" for their analysis. "What was the most important scene?" she asked.

> *Girls:* Hey, guys, say something. Why the questions are always addressed to us? We have already answered.
>
> *Riku:* The message of the film [reads from the questionnaire]: believe in your dreams, little boy. People should do what they really want to do. And now, shall we watch *Wallace and Gromit?*

After some arguing, the teacher decided to start the movie. For students, the classroom and school space was more a home with familiar devices like DVD players, televisions, computers, video and still cameras, and of course their own devices that they brought from their real homes. The feeling of home was partly created by creating unofficial spaces inside the school and partly by having trustworthy adults and peers around. What the visual arts teacher described as copying from action movies was from my point of view partly a process of domesticating the official school space to a more livable space and obtaining some strong emotional pleasure in schooling. This kind of tactic is in a way a practice of making something more homelike. As a pedagogical institute, a school is far from home, marked with timetables, routines, orders, academic assignments, and locked doors. Watching and especially making films in special locations in schools opens a new space for an embodied action. Students also make the space with their bodies and with playful fights, dance, and running in the school space.

The movie course allowed students to go outside the school to shoot and edit their films. At the end of the course, they visited the municipal Media School, which was equipped with a studio and multiple-camera setup, along with recording and editing capabilities. Most of the students were excited to be able to shoot their stories with professional cameras and use the studio system. As a good example of a tactic, during the end of the course one group decided to make a short film of school meals with the statement that in McDonald's, they can get better food than in the school. The video was entitled *McDonald's vs. School Food.* The group shot the video in the school canteen and in McDonald's, which unfortunately caught fire just after their shooting. The group probably had the last images of McDonald's before the fire.

McDonald's as a site for their film and the statement about the school food are readable as a resistance toward official school spaces. Another group started to shoot in the Media School a story about Facebook, exploring its positive and negative aspects. Once again, students chose a theme from their everyday life out of school. Gomi was also making her video, this time totally independently and out of school spaces. Nobody knew, the teacher included, what Gomi was really doing. She made her video at home after school. The outcome was some kind of surprise. She planned to do a film of public transportation, but the outcome commented on the school. In her video, we can see a girl starting to sleep until she hears a slam. She thinks that it is a ghost of the previous owner of the house who has come to punish her. In the end, the girl says, "or then it is my history textbook." In this film, the history book causes some nightmares, and the school has arrived to the home space as a ghost.

Animation Workshop in Visual Arts Class

As seen already, creative media production is part of everyday life of young people. The visual arts teacher reasoned that because students did so many different things at home, they were not self-evidently interested in doing the same things at school, for example, making a short animated film, which has been an assignment for eighth-grade students in visual arts for years. The animation course needed plenty of time and took several days. The teacher thought that students wanted to make the animation a little bit faster than before:

> It's not a dramatic change, but I have noticed during these years that students would like to get ready a little bit faster, so that their motivation will stay. And the problem is in the works like animation that need time. Or there is some kind of contradiction: is there any sense just to do short projects and clips and not go deeply to the core of animation. It may stay superficial. (Interview with the visual arts teacher, male)

According to the teacher, the problem is not in analysis or motivation but in the need for students to concentrate for a long period of time. Other teachers also said that making media in the official school context is sometimes difficult. Students do not always come up with good ideas about what to do, or they spend time arguing in the working groups.

Moreover, so-called defensive teaching and learning in traditional schooling may affect that students more or less pretend inability. In their ethnographic study in U.S. high schools, Garrison and Bromley (2004) found that students sometimes acted as though they were unable or unwilling to do computer assignments. This was because teachers restricted and controlled students every single act, by telling them what to do and when. Defensive teaching expands control beyond what is necessary and stifles motivation for learning. Students may resist excessive authority over how they learn at school and accomplish this in different ways, for example, by "acting busy" and feigning inability. This is an example of students' general resistance to schooling by using school technology in the wrong way or showing total unwillingness. They just do not want to follow teachers' structured ways of guidance by asking them to follow step-by-step directions. Students prefer to do their own thing and do it outside the school if the school does not offer opportunities for that.

But teachers have specific goals they want students to reach. In an animation course, students are supposed to learn basic technical skills for animation production. The visual arts teacher emphasized that the goal of visual arts is to teach media as a means to produce something. The difference between mother tongue and visual arts, according to the teachers, is that in mother tongue, students learn more media literacy and about the media. The core curriculum for visual arts had some technological emphasis, but in the school, the resources were limited:

> We had plans that we can get one video camera for example every other year, four per visual arts class all together. In other words, eight camcorders. If we have 26 students each it is at least minimum, so that we have four camcorders per class. So, that we are able to do at least something. But nothing happens. (Interview with the visual arts teacher, male)

The animation course was conducted with three camcorders and laptops using iMotion software. These camcorders were all both visual arts classes had together. ICT and mother tongue also had one camcorder each. Because of the lack of camcorders, the group sizes in the animation course were quite large, about eight students per camera. The teachers said that some years they have had more camcorders and the work has been much easier and faster.

One laptop in the class was reserved for music and effects production. Making animation was a school-curriculum-based media production that was not possible to do outside of school because of the "studio-like" feeling. Students did not have the same software at home and could not move files from school laptops to home computers in order to do out-of-school work.

The idea of the animation course was metamorphosis without any specific content. Metamorphosis is a unique form of animation, literally changing an image into a completely different image, through the evolution of the line. Metamorphosis creates fluid linkages of images through the process of animation itself without editing (Wells, 1998, p. 69). For example, in the animation films produced in the class, we see a ball's transformation into a balloon, kite, and airship. Interestingly, the evolution of the line can constitute narrative constructions, and students seemed to be conscious of these possibilities.

When I arrived at the class, all students were working. Some were drawing images, some were filming an animation, and some were recording sound effects.

Figure 5.2: Shooting an Animation in the Visual Arts Class

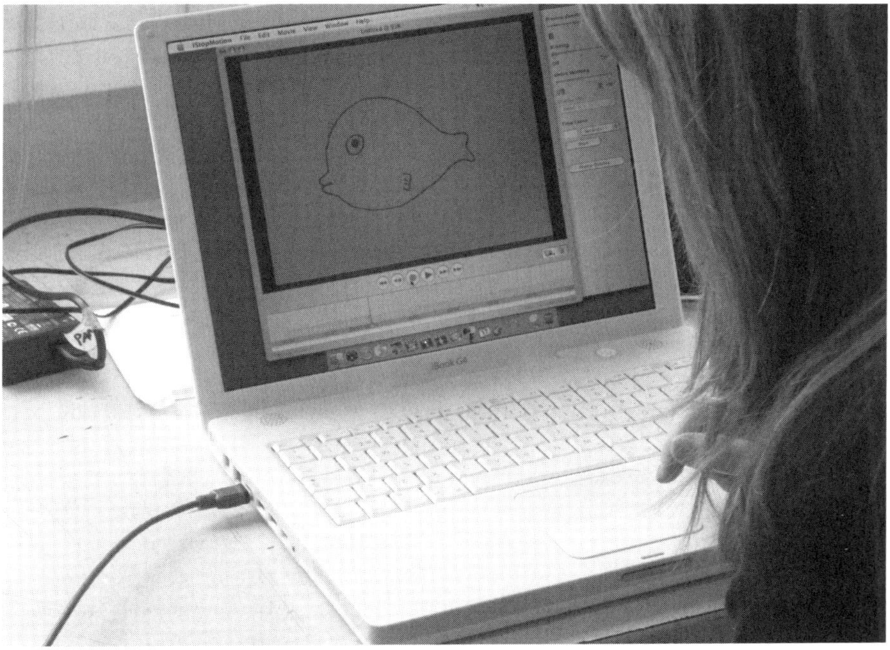

The teacher explained for one group how to use GarageBand software for recording and editing sounds:

> We do not choose *Software* but a *Real Instrument* from the menu bar. If we would have a guitar we could plug it in but now we are only speaking and we choose *Vocal*. I choose here *No Effects*. We could choose something that changes the sound a little bit. By clicking *Create* we create a track. After that we are able to use the track. The microphone is here [pointing to the inbuilt microphone]. We can speak, beat something or do anything else. Whatever you need.

Students were listening and the teacher took a piece of paper and crumpled it in front of the microphone. Somebody asked how they were able to transfer the sound to their animation film. The teacher explained that the sound has first to be transferred to the editing software FinalCut:

> In FinalCut we can edit your sound or music and put in the track in the place where you want it to be. You can edit it afterwards, as you will. But I'll give you some advice; try to put it roughly in the right place when you are record-

ing and watching your film. Another group did so by having two laptops side by side and they watched where they would need a special sound like a slam.

After this short guidance, the group was immediately ready to record sounds. Students tried to simulate the sound of fencing using chisels they found in the classroom.

As the teacher said, students were practicing new software and means of animation film. They were not familiar with IStopMotion, GarageBand, or FinalCut software, but they quickly determined how to use the software for their means. Rebekah Willet (2009, p. 18) argued that these kinds of software scaffold learning. Learning the software is usually intuitive. Different software packages have the same kind of logic and menu bars with options and young people are used to learning by the method of trial and error.

In some way, the learning was similar to learning with video games. Gee (2007) outlined several different learning principles that are inherent in gaming: for example, learning through gaming is active, meaningful, multimodal, scaffolded, and entails participation. He argued that rather than through abstract exercises and textbook reading, learning happens in the context of an activity and is therefore more effective. However, the students were not alone with the new software and animation techniques. The teacher gave step-by-step instructions whenever needed and helped students solve problems. In fact, learning during the animation course was a combination of teacher's instructions, trial and error, technology-enhanced scaffolding, and peer-based learning. Working in groups requires communication between the students and simulates the way they learn outside school with each other in different networks.

During the animation course, the metamorphosis of the line was not the only metamorphosis. At the same time, we saw a metamorphosis of the official school space and classroom to an unofficial animation studio. Students worked freely in groups and the teacher gave some instruction if students needed it. Students who were drawing pictures for an animation listened to music and kept their mobile phones at their desk ready for text messaging and other communication with peers. From time to time, a few of the students asked the whole class to be silent while they were recording sound effects. Others respected this request commendably.

Visual arts is a school subject where frontal teaching is exceptional. Usually, students are working alone or in groups after the teacher's instructions. The classroom is not the only space for learning. Students visit exhibitions, draw sketches outside the school building, and use the whole school building for different assignments. Students work quite independently and sometimes do different tasks. Somebody may not have finished earlier work when others have already started something new. This is sometimes a problem, as is the classroom itself. The teacher said that the classroom is not suitable for visual arts teaching. They did not have enough space to store students' work or do any project with materials that need more space. Like in other classrooms, school desks were in rows and lines. They were easy to move, but the space was not easy to manage so that almost 30 students would have a good work space.

ICT Teaching

Some projects in the visual arts class were done jointly with the ICT class. The school has organized ICT teaching in a special way. It was integrated into other school subjects like visual arts, history, mother tongue, and foreign language. In Finnish schools, ICT is not an official school subject and therefore schools had to find different solutions for how to teach it. In some schools, ICT is taught as an elective subject.

Joint teaching at school needs resources because two teachers are in the classroom at the same time. Usually, teaching hours are clearly divided in Finnish schools so that it is not possible to have two teachers in the same class simultaneously. But there are also some advantages to doing this. One goal in the Maple School has been that subject teachers will become more courageous and skilled in using ICT in their teaching. As mentioned before, subject teachers in Finland are not especially keen in using digital technology in teaching. According to the international SITES report (Kankaanranta & Puhakka, 2008), about half of the math teachers have used ICT in the class. Among 22 participating countries, Finland is ranked 10th in using ICT in the classroom.

Teaching ICT with other teachers in the Maple School was an attempt to help them find ways to use ICT. For example, in the English language class, students learned to use PowerPoint software with ICT and the English teachers. They had an assignment to create a PowerPoint presentation about their virtual trip to the United States and write slides using English. One of the girls asked if it was possible to send the presentation to herself via e-mail, so that she could continue the assignment at home. Once again, the assignment was transferring the boundaries of school and home and bringing school literacies to the home.

When I was observing the teaching, it was a hot, early autumn day. The computer lab was in the school's loft without a window. The computer lab included 15 computers but the air conditioning was broken. The computers were lined up in rows in the same way as in many schools that I have seen. Computer labs are a traditional way to organize ICT teaching, but they are not necessarily the only and most effective way. Gavriel Salomon (1990) wrote as early as 1990 that separated computer labs may reduce the use of ICT in basic learning activities in classrooms. In the Maple School, where most classrooms had only one computer, the two computer labs were the only possibility of giving students access to the computers.

The class started in a traditional way: the teacher introduced the software and assignments. After that, students started to find information and images using Google and Wikipedia. This was a common way to find information for students' own presentations. I asked if the PowerPoint software was easy for them to use. Nobody denied this and some of the students said that they had used it at home as well. When the students were designing their PowerPoint slides, the teacher showed me some earlier presentations. Some students had also integrated video clips into their presentations. In the interview, I asked the ICT teacher why they studied PowerPoint instead of weblogs as a kind of medium to present a trip: "They learn themselves those things when they do and surf on the Net. I don't get any pressures because they find these new things like blogs, Facebook and MS Messenger themselves."

PowerPoint and other Microsoft Office applications dominate the official school space and teaching. Neil Selwyn (2011) noted that cutting and pasting online material retrieved from Google and other search engines into Microsoft Word documents and PowerPoint presentations in fact are the most common practices for teaching ICT. However, there are some problems related to Microsoft Office applications. They are developed for the business sector and not for schools, and they mirror the logic of the one-way distribution of information valued in the business world. School administration and some teachers have argued that these kinds of applications are most important to learn and that students do not have the necessary skills to use them. Teachers are familiar with this kind of technology because of their university years and academic studies where PowerPoint slides and Word documents were used almost daily. Learning to create presentations and documents makes sense from this perspective. In fact, some students in Finland have problems with presentations and word-processing programs or with e-mails and attachments because they are not used to using these kinds of software, which are not necessarily taught at the school. The lack of ICT teaching concerns many, but the teaching of ICT programs as such is insufficient without the perspective of broader media education and an understanding of the media practices of the youth.

By the end of the class, it was hard to get fresh air. Some students began to feel uneasy. Some opened Facebook sites. Someone even shouted to the teacher that he had sent a friend request to the ICT teacher on Facebook. When the ICT teacher moved away from the classroom, the English teacher started to feel uncomfortable. She told students that she was not able to help with the PowerPoint software because of her lack of skills. In this case, joint teaching did not encourage the subject teacher to use digital technology in the class; on the contrary, she pushed the responsibility for this to the ICT teacher.

Several other joint teaching cases showed that both of the teachers were able to help students with the software. As there are differences among students, there are differences among teachers as well.

In the seventh-grade joint class of history and ICT, the assignment was to create a home page for some chosen Finnish historical person. The ICT teacher advised students to use image processing in order to get images for the web site. The computer lab had a new image processing program, Adobe Photoshop Elements, instead of PaintShop Pro that was installed on previous computers. The ICT teacher was not familiar with the new program, which had different logic than PaintShop Pro. When students asked something about the software, the teacher tried to study how it worked. The history teacher also helped when students had problems with image processing and commented, for example, on how to scale images. He was interested in ICT technology in a different way than the English teacher.

Girls in one group chose Finnish writer and social activist Minna Canth (1844–1897) for their assignment. They found some contradictory information on the Internet about Canth's husband. The history teacher explained that the newer information from the Internet is probably more correct than the older information from the book that they found in the reference library in the history class. "This is quite normal in these kinds of projects that students find contradictory information," the teacher said. Interestingly, the assignment led students to understand that information from different sources can vary. The teacher noted that, "usually there is no right answer because we do not have such kind of authority that could tell the latest information."

Students used their information literacy and compared different information and sources, and what is even more important is that they understood that information is created by somebody somewhere and that there could be contradictory information. Information does not fall from the sky, as it may be understood when teachers transfer information via the pedagogical "banking system" method. Now the students had the chance to search for information by themselves. Juha Suoranta and Tere Vadén (2010, p. 70) pointed out that some Internet sources, especially Wikipedia, are revolutionary information sources because they offer an epistemological shift: "edit" and "history" buttons in the Wikipedia page indicate that the knowledge has a history, it comes with a past and a future, and it is not black and white. The history teacher allowed students to use Wikipedia as a source for their assignment and build knowledge in an alternative way than in traditional schooling settings.

Searching for information and producing a web page was also an example of digital literacies that are needed in remixing practices. The students gathered information, images, and other multimodal texts from different sources and composed their own content. The teaching and learning situation was also different;

students in a computer class with two teachers admitted their own lack of technical skills because of the new programs installed in the computer lab. The web page editor, called Nvu, was new to the teachers and students. Some students said that they had designed web pages before using HTML language. One boy told me that he had designed a discussion forum for a game site by writing HTML code with Notepad.

The ICT teacher emphasized that downloading images in order to use them on a published web page was not allowed due to copyright laws. She gave an example of a student who downloaded images of dogs from the Internet and linked them to her school web site. The copyright owner discovered this and demanded some indemnities. The school managed the situation with an apology and by taking the images off the page. When the teacher asked if anybody knew sites where students could download images freely, one girl mentioned Flickr: "You can choose free images from there...it is something "copyrights common" in English...." The girl meant a Creative Commons license, which allows users to share their images without full copyright. This was unknown to the teacher but she gave some other tips: "I quite often use the key words *clip art* or *clip art free* or *gifs*. Let's try clip art. Usually you can find free images and you can use them."

Interestingly, the joint history and ICT class moved from web design via image processing to issues of copyright and Creative Commons, information seeking, and knowledge building. These are all issues that are important from the perspective of creative media practices and media and digital literacies. Students learn these issues by doing and through collaboration with others and teachers. The teachers do not offer a role of all-knowing adults but colleagues who also need information and do not always have the "right" and final answers. Elisabeth Soep and Vivian Chávez (2010) spoke about collegial pedagogy, which creates a joint learning environment where both parties, students and teachers, may deepen their learning experience: "Under collegial pedagogy, young people and adults actually make work together, revealing their investments and vulnerabilities to one another in concrete ways" (p. 55). (We talk more about collegial pedagogy in Chapter 6.) Both parties are free to show that they do not know everything and that they need each other in order to find solutions and make media productions. Some curriculum-based media productions in the Maple School demonstrated this in practice and gave the students more room to use their skills and competences in the classroom.

Media Production in History and Civic Class

> School has to compete with other media.... If young people play games and hang out on Facebook or something like that, their reality is so full of stimulation. They get new and exciting things every moment. At school this means concretely that we have to

> work hard in order to keep students tuned, so that stimulus in school correspond to their reality. If you think that when the teacher is talking and nearly 30 people are listening, it is in a way a really passive situation. How can a student react? They are silent and write and ask questions one at a time. School reality does not correspond to that in their free time they can be agents all the time and decide what they do. I think that contrast is huge. (Interview with the history and social science teacher 1, male)

The history and social science teacher emphasizes here the *agency* of students, which is not self-evident in official school settings. He recognizes the agency that young people have outside the school, especially in their own media practices. In an interview, he said that he tries to use different pedagogy than pure frontal teaching with questions and answer sessions. He saw the gap between the traditional school and students' culture and tried to "awake a will to listen and learn" with visual pedagogy where he used PowerPoint presentations, maps, animations, photographs, and clips from the Finnish Broadcasting Company (YLE) TV archive.

> I try to activate students that it is more than knowledge, sometimes experiences or something like that. But it could not always be something that "hey, now we play games" or something like that. If I could, I always try to find some way to regenerate. In history class, I sometimes even try to develop some games, for example I have explained the risks and problems of investing in stocks with a help of a game. (Interview with the history and social science teacher 1, male)

Both history and social science teachers at the school had an open attitude toward new and partly experimental teaching methods. They gave students the opportunity to use their media practices and, for example, to make videos and animations instead of written essays. As I mentioned before, Gomi was one of the students who used her gifts and made visual narratives for the history and social science class. In seventh grade, she made an animation film about the 1904 assassination of Russian Major General Nikolay Bobrikov in Finland. In ninth grade, she made a short film for a social science class.

The social science students had a "life career assignment" where they were free to use any kind of content and material from popular culture. Social science education in Finland is based on the different branches of social sciences: political science, economy, social policy, sociology, and law (Virta & Yli-Panula, 2012). In education, these are related to content as well as an understanding of basic concepts. In the practice of teaching, attention is given to society both at the macrolevel (parliament, the government, voting, economics, law, municipal services) and at the microlevel and to how young people encounter social decision making and economic issues in their everyday life (Virta & Yli-Panula, 2012, p. 194). The assignment was targeted at the microlevel. Students had to present the life history of a fictive person within society, where different normative situations are based on law. Students had to learn, for example, about different legal issues that come into play when one is considering birth of the child, compulsory education, marriage, divorce, child and housing benefits, and so forth. Students used different media platforms and social networking sites to present different life histories, for example, magazines, animations, children's books, posters, photographs, Facebook sites, Habbo, and so forth. The students were seemingly pleased to have invented something the others had not. The teacher explained this assignment in the following way:

> The starting point was that first issues in social science are fragmented but they are related to each other.... We speak about autonomous learning. I see that a student really has to deal with information and find information, understand it, and make a big picture of it. A student has to be active and go through these things.... This is especially true for students who underachieve in tests or just do not learn from books or can't write proper answers. Many students that are not so good at tests are good at this. (Interview with the social science teacher 2, male)

The teacher added that students work so hard it is almost cruel to assess their works in the school's traditional manner. He mentioned that "in the beginning working is just a mess and it seems that they can't do anything, but when the due date comes, they bring really fine works." Students do almost all the work out of school with their own devices and home computers. Some students, like Gomi, even published their productions on the Internet on their own social networking sites.

The assignment is a classical example of the "learning by doing" pedagogy which, of course, has its own challenges: it takes time and space. Students may find it difficult to receive personal guidance from the teacher because there are so many students in the classroom, and the teacher can find it difficult to assess students' very different presentations.

Another "learning by doing" assignment I observed was a newspaper project in a seventh-grade history class. Newspaper is a classical learning material in Finnish schools, as we have already seen. Actually, Finnish newspapers have almost 50 years' experience in cooperating with schools. The Finnish Newspaper Association has arranged a special Newspaper Week every February since 1994. During Newspaper Week, schools can order local newspapers for free for every student. The Finnish Newspaper Association also publishes pedagogical material for teachers in order to lower the thresholds for adaption of newspapers as a part of a school day. In recent years, newspapers have also taken young readers into account by publishing news and reports during the Newspaper Week that are targeted especially for the children and youth. Some newspapers have also used young people as journalists.

In 2010, the mother tongue teachers conducted a newspaper project during the Newspaper Week. Newspapers have been used at other times in classes as well. In history class, the assignment was to design headlines and a front page from 19th-century historical events, such as the Finnish War (1808–1809), the Diet of Porvoo (1809), the Grand Principality of Finland (1809–1917), or the

Figure 5.3: Students Are Designing Old News to a New Context

Congress of Vienna (1814–1815). Students had to find the most interesting and important information about the historical events and write an article using the genre of short news. The teacher commented that "students are used to hands-on projects. Why not utilize their skills?"

Seventh-grade students were eager to start working on their "own newspaper" and most groups started to design the front page visually and by drawing images. Some spent a lot of time trying to write using Fraktur. The teacher showed examples of fonts that were used in old newspapers. Even though the fonts were old in students' work, the journalistic style was new. For example, the *Aamulehti* that students "edited" had learned that there was a scandal in the Congress of Vienna (Figure 5.3). The headline was as follows: "In a dancing congress people danced dirty way." Students played with the words "sääty" (estate) and "säädytön" (dirty/out-of-estate). Another newspaper published a Gallup Poll: "Do you want the regime of Russia? Answer by post." Students used a lot of humor and tried to find clever historical jokes.

Once again, the assignment required space and time that was not suitable for traditional school settings with 45-minute sequences and small school tables. Learning by doing is not simply a task to be implemented in official school spaces; it can provide a personal learning process. It goes far beyond a simple information transfer and contextualizes the meaning of subject matter in a new way.

The assignment in the history class was a kind of remixing process. It restructured the context of content in a similar way as remixing shifts any context, like adding comedy music to a horror film. This kind of work puts meaning to the contextualization, where learning is focused not so much on an interpretation of *what* something means but *where* it has meaning and how it changes (Thomas & Brown, 2011). Thomas and Brown (2011) wrote that this kind of learning is an act of imaginative play, the "how" of information that is needed, especially in the age of the Internet and digital culture. We need the skills to find and evaluate information in different contexts where no guarantee of the truth of the information is given. Meaning can be manipulated in different contexts and we have to gain awareness of the play and possible reconstructions of the context.

· 6 ·

LEARNING AND MEDIA LITERACIES

Civic Participation

At Maple School, some teachers tried to give students a voice and opportunities to express themselves with different media. For example, at the end of my ethnography, I observed a project where students produced a short video for a competition entitled "Mix a Minute." The competition was introduced by local media pedagogue Jari who visited a few times in the movie course. The aim for the competition was to express a personal opinion with a 1-minute video film. Jari wrote instructions on the classroom whiteboard:

Mix a minute

- in my opinion...
- two different points of view from the same issue (From my notebook)

At first, students did not understand the instructions. Jari explained that the 1-minute video could contain all possible visual material from still photos to animation. The goal of the project was for the video to "address two different sides of the same thing," as Jari explained. He gave examples of "two sides: public and private, slow and quick, beautiful and ugly, whatever opposites." Students were required to use different audiovisual means and have an original idea. Jari mentioned that the background of the competition were bulletins that were common in Finnish television, especially in the 1970s and 1980s. In short television spots, bulletins tried to warn people about everyday dangers like traffic accidents, intoxicants, and fragile ice. Jari asked if anybody had ever seen any bulletins on television, but the students remained silent. There was a clear generational gap between the students and the teacher. They did not have the same everyday experience, which is a common problem in media education. Buckingham (2003a, pp. 323–324) pointed out that in media education, students need to bring forth their own tastes and emotions instead of simply being given instructions and experiences by teachers.

This example from the movie course shows that teachers try to provide students with opportunities to tell their own stories and have a voice in the class, but the initiative and structure comes from the teachers. This is generally a major problem of education, known as "the pedagogical paradox." It was introduced by Immanuel Kant in 1803 (Kant, 1900). In general terms, we celebrate that young people have a free voice and are capable of autonomous judgments, but we take for granted that teachers determine how students should practice that autonomy. For example, in the 1-minute video project, teachers clearly considered students to be political agents but in the formulated structure of the assignment. Buckingham (2000, p. 168) pointed out that traditionally, young people have been prevented from having any political influence and involvement in the political discussion concerning even their own rights. Times have changed and modern society calls for political activity and active citizenship as a part of the everyday life of youth. But the active practices may be different than teachers have expected.

When students started to plan their ideas for their 1-minute video, Jari gave examples from scratch video as a critique of mass media and commercial culture. He explained how a scratch video changes the original meaning of the media text and highlighted it as a possible starting point for a 1-minute video. The idea of the competition and the assignment in the classroom was clearly to take a critical perspective to the mass culture and society. The visual arts teacher tried to encourage students to find anything that annoyed them, "something in which it is possible to intervene or have an influence." Students tried but were not able to find any good ideas for their videos. After class, the visual arts teacher made a notion that "boys play [video games] and girls make up." She suggested that in their teenage years, students are not very involved in active participation.

In an interview, one of the social science teachers said that, from his experience, 14- and 15-year-old students are not yet involved with political activism, although that they have some interest in social issues:

> They may be interested in different issues and talk about them and ask questions. But I do not know that anybody [in the Maple School] is involved with any political activ-

> ity or anything like that. I think that it is a question of their age, later in upper secondary school they wake up to ask "what can we do." But now in grade 9 when in social science class we talk about different possibilities to have an influence in the society they sort of wake up and think that "yes, we can collect names for the petition" or "hey, we can go to a rally" or "wait, we can have an influence on these and these things."... But they do not have any means or motivation to change things.... When there is time concretely to do something, there is nobody who will do anything. They have a wish that "we want to make decisions" but when they are supposed to do something, they want somebody else to put it into practice. Their activity is part of their will to power, but not to their feeling of responsibility. And for me, it is a little bit silly. (Interview with the social science teacher 2, male)

I must add that active participation and critical thinking are not a strong part of school culture in Finland. Finland is reported to be one of the countries where students' interest in politics and citizen activity is lowest among 38 countries around the world (Schulz, Ainley, Fraillon, Kerr, & Losito, 2010, p. 51; Torney-Purta, Lehmann, Oswald, & Schulz, 2001, p. 117). Finnish students are knowledgeable about society but not so much regarding "civic attitudes" to participate and change things. Young people seem to be more passive consumers than active participants in society. According to 2009's Civic and Citizenship Education Study (ICCS), Finnish eighth graders were ranked among the lowest of similar students in other countries regarding an interest in politics. Young Finns think that they do not have an influence in school, especially regarding the school timetable, teaching materials, school regulations, and contents of teaching. They think that they can influence society via the Internet and their social networks (Opetushallitus, 2011; Suoninen, Kupari, & Törmäkangas, 2010). If schools do not provide opportunities for active participation, young people try to make their own room for active citizenship and participation by using digital media and creating their own heterotopic spaces, as we have seen.

Sonia Livingstone (2004) wrote that media literacy includes a promise "that it can form part of a strategy to reposition the media user—from passive to active, from recipient to participant, from consumer to citizen" (p. 20). This is one reason why media literacy has been incorporated into communication and education policies, for example, in the European Union level and at the national level in many countries. Media literacy is a key term when it comes to participation in society. But "participation" is not a self-evident concept even in the age where new media is celebrated for its participatory potential.

Usually participation is seen from the point of view of democracy and politics. Nico Carpentier (2011) made a distinction between "minimalist democratic participation" and "maximalist democratic participation." Minimalist participation tends to focus on representation of power and macroparticipation, which means participation in the whole society and public sphere. It usually has quite a narrow definition in politics as an institutional business of the elite and voting.

Maximalist participation varies from macroparticipation to microparticipation, which means participation in the spheres of school, family, workplace, church, and community. It has a broad understanding of the political life as a dimension of the social. Participation is also sometimes divided into full and partial and authentic and pseudo (Carpentier & de Cleen, 2008). Full participation is seen as a process where all individual members have equal power in decision making. In partial participation, individuals influence each other, but the final power in decision making rests with one party only. Political institutions, for example, offer people a chance to express their opinion in decisions that affect them, but the power of decision making is in the hands of politicians and other officers.

From this perspective, participation has certain limitations and, for example, participation on the Internet usually offers more pseudoparticipation, more just access and interaction than participation in the strong and full sense of the word. Participation online in this sense usually also has an organizational and commercial context, where structural control is in the hands of firms and professionals (see Carpentier, 2010). In the democratization process through the media, models of full and authentic participation are "extremely difficult to put into practice," as Carpenter (2010, p. 55) argued. However, we have some evidence that participation may be successful and authentic, for example in the case of the Arab Spring, where demonstrations and protests occurred in the Arab world partially by using social media as a way to organize protests, communicate, and raise awareness.

Nick Couldry (2008) argued that neoliberal democracies suffer from a "voice crisis" due to the market-driven politics where political and social organizations have only limited influence over economic policy. At the same time, political organizations are so market-friendly that almost all political settings have been done from the perspective of economics: "The result of market-driven politics is that the national democratic system becomes less able, even in principle, to deliver voice" (p. 20).

But voice matters as a dimension of human life, as a process of making meaning. Participation can also be understood more as communication and engagement with fellow citizens at the microlevel and not so much in political decision making at the macrolevel. Beyond the democratic theory, participation is in relation to the cultural and symbolic realm of societies and the politics of representations (Carpentier, 2011; Hartmann, 2010). This dimension of participation is adopted by the students when they communicate, make meaning, and build their identities online, for example by writing lifestyle blogs. Young people participate in the media and in the microlevel of everyday life by creating content.

From the perspective of school and learning settings, the starting point is more democratic-political approaches to participation. In the action plan of "participation supporting media education" by the Finnish National Board of Education (2011), participation is defined as listening to the voice of children and young people in the school culture and all procedures and decisions regarding the school (p. 7). Participation is understood here, first of all, as microparticipation and partial participation, but it enlarges according to the action plan to more macrolevel participation in the whole society. At the curriculum level, media education that supports participation is conducted mostly in cross-curricular themes of participatory citizenship, entrepreneurship, and media skills and communication. The goal of cross-curricular themes of participatory citizenship and entrepreneurship is, among other things, to help students develop capabilities that are needed for civic involvement. However, the action plan takes students' questions and need for involvement into account in relation to their lifeworld and everyday life. The action plan emphasizes sociability, collaboration, and multimodality as a starting point for media education. Moreover, it says that the media of the youth outside of the school has to be considered alongside traditional learning environments at school. In this way, the action plan creates a bridge between out-of-school literacies and school literacies, highlighting the participatory dimension of out-of-school media practices.

Participants in Participatory Culture?

My survey (N = 305), after all, suggested that almost half of the students are inclined to have an influence on the Internet. It is thought that the Internet and

Figure 6.1. Importance of Activities on the Internet ($N = 305$)

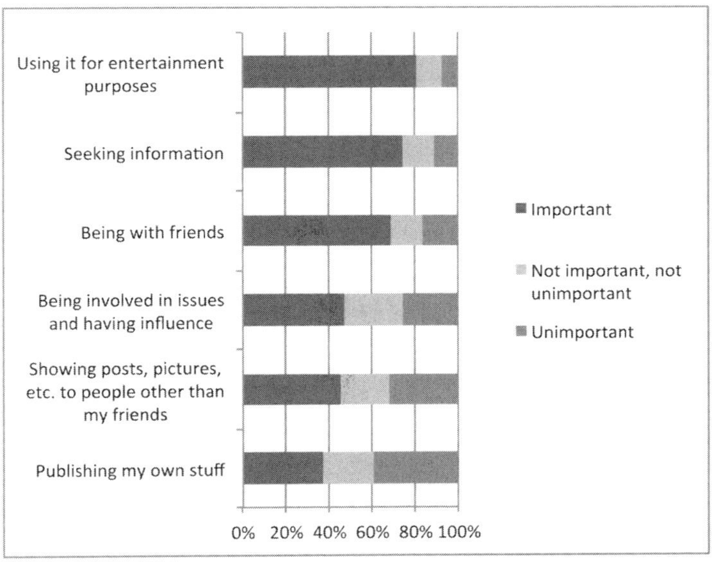

especially Web 2.0 can create a new form of political activism. Natalie Fenton (2010, p. 19) explained that the Internet is considered to raise people's awareness, give voice to those who do not have one, offer social empowerment, and be used as a tool for social change.

In my survey, I asked students to list their most important activities on the Internet. For 47%, having some kind of influence was important, and for 39%, it was important to publish and share their own material. Compare that to the responses to the question, "how important is it to enjoy the Internet (listening to music/watching movies/surfing)," where 81% answered that enjoyment is important. According to Ito and her colleagues (2010, p. 37), the vast majority of young people in the United States desire to "hang around, meet friends, just be" on the Internet. Even in China, young people's Internet usage seems to be "overwhelmingly apolitical, with a focus on entertainment, lifestyles and private issues" (Liu, 2011, p. 177). From this point of view, arguments about a new participatory culture where young people use the Internet as active citizens for their self-expression and creative practices seem to be without support on a large scale. But if almost half of the school students say that the Internet is important for some kind of activism, participation culture seems to have an important role in their lives. What has to be considered, however, is what kind of participation or influence students want to have and what they consider meaningful.

Participatory culture is a term coined by Henry Jenkins that means new media ecology, where creators can develop their voices and identities through interaction with engaged peers and audiences (Jenkins et al., p. 2006). What is thought to be new with participatory culture is the possibility for people not only to be media consumers but to be active media producers or pro-ams (pro-amateurs). For Don Tapscott and Anthony D. Williams (2006), participatory culture implies a promise of collaboration through which peer production "will harness human skill, ingenuity and intelligence more efficiently and effectively than anything we have witnessed before" (p. 18).

The Internet creates possibilities for young people to create and share their productions easily. Nevertheless, the majority of young people prefer media consumption and content-based activities to media production. EU Kids Online research suggested that only a quarter of children and young people across the Europe reach the most advanced and creative step of the "ladder of opportunities" (Livingstone et al., 2011b, p. 14), which includes young people creating and sharing different things on the Internet, like writing blogs, spending time in virtual worlds, and so forth. Almost all children use the Internet for schoolwork and to play games alone but fewer youngsters go up to the next steps of the ladder of opportunities. Half of the children and young people reach the fourth step, where they download films and music, share content peer-to-peer, and play games with others online. Next, the aforementioned creative step is the hardest to take.

There is, however, a minority of young people who want to create and share their own media productions. My study shows that young people who are engaged with creative media production use the Internet for different purposes. In a sense, their Internet use is individualized. Manuel Castells (1996) wrote that in the digital world of networks, individuals constantly redefine their lifestyles and consumption patterns. Media experiences and the subjects of interest are diverse.

We do not need to underestimate participatory culture or its significance. It has potentiality to change culture and media consumption and give voice to the people. Axel Bruns has introduced the term *produsage* (2008) as a model for describing RW culture and user-led content creation online environments. From this point of view, participants of online environments are at the same time users and producers of information and knowledge. Bruns emphasized that produsage is a model away from the industrial producer-distributor-consumer chain to a postindustrial hybrid model where usage is necessarily also productive (p. 3). Produsage is described through four key characteristics: (a) a broad-based, distributed generation of content by a wide community of participants; (b) fluid switches of producers between different roles as leaders, participants, and users of content; (c) a generation of information always unfinished and continually under development; and (d) a location within a framework of permissive regimes of engagement. What is needed is a set of skills in order to be produsage and fully participate in the production-usage-interaction. According to Bruns, these are (a) creative, (b) collaborative, (c) critical, (d) combinatory, and (e) communication skills. This framework also opens a need for a new kind of learning environment that is not a ready-made learning environment with established teacher-student roles but a new emerging space that grows along with produsage. Learning is taking place in everyday media practices when young people read and write media with personal needs and actions and with special capacities.

The Voice of Students

Produsage requires skills that may help people use their voice and participate authentically and fully in order to have an influence and empower themselves. Empowerment is one of the most important objectives of media education. Empowerment became an educational buzzword in the mid-1980s, but it still has an important role in relation to literacies and media literacy. It is also significant when we are talking about the importance of people's own voice. Empowerment is a difficult concept and has rarely been given an adequate definition (Lankshear, Gee, Knobel, & Searle, 2002, p. 63). What is common with many approaches to empowerment is that it is some kind of inner feeling of power and ability to do things and participate. It is close to the concept of emancipation by Paulo Freire. Freire used the method of dialogue in order to enhance people's ability to use their own voice and not just reproduce what is already said. When we are talking about media literacy, an empowered agent is an active producer of media representations and is able to understand and use media rhetoric and different effective means of media language and style.

Empowerment is a relational concept in the same way as the concept "freedom" (Lankshear et al., 2002, p. 65). Traditionally, there has been a difference between positive and negative freedom. Positive freedom means "freedom for what," what people are free to do, be, or have. Negative freedom is a "freedom from what," from what people are free. In the same way, an empowerment is from something to something. In fact, it involves at least four variables (Lankshear et al., 2002, p. 65): (a) the subject of empowerment, (b) structures in relation to which and from a person is empowered, (c) the processes through or by which empowerment occurs, and (d) the outcomes of the empowerment.

Lankshear et al. (2002) noted the following:

> In other words, claims about empowerment should spell out the contents of A-D in the following schema:
>
> A (the subject) is empowered in respect of *B* (some aspect of the discursive structuring of power) by/through C (a process or quality) such that *D* (some valued ends or outcomes) may—and ideally *will*—result. (p. 65)

From my perspective, an empowered person is able to use his or her own voice in order to have an influence and recognition for the worth of his or her contribution. Empowerment and voice are questions of civic participation and participatory culture, possibilities to be part of democratic society.

Voice is obviously one of the crucial aspects of the legitimacy of modern democracies. Treating people as if they lack the capacity of effective voice is to treat them as if they are not human beings (Couldry, 2010, p. 1). For example, for French philosopher Jacques Rancière, an act of speech is something that qualifies who is able to think and participate in society. Traditionally, slaves, women, children, colonized people, and the poor have been without voices, and we have heard only their sounds of hunger, hysteria, and cries (Rancière, 2004, p. 5).

The question of voice is in relation to the question of equality. This is a very well-known idea in the sphere of politics, where voice is usually understood as expression of opinion or more broadly "as the expression of a distinctive perspective on the world that needs to be acknowledged" (Couldry, 2010, p. 1). In *Politics*, Aristotle makes a well-known distinction between *mere* meaningless *sound* (*phoné*) and *voice* or *speech* (*logos*). Only the latter is the medium of political deliberation and action, the former being the capacity of communicating basic sensations of pain and the like. In modern democratic societies, we celebrate "voice"; we are looking for different kinds of public spheres where everybody is allowed and invited to express their opinions. In a school pedagogue, we believe in dialogical methods instead of hierarchical "banking" and teacher-driven learning as well.

We also believe in the value of emancipation, where teachers let students have their own voice in a classroom and become agents of their own life experiences. But it is also criticized that such kind of emancipatory process is always something that is *done* by somebody (i.e., teacher) to somebody (i.e., student) and the process is therefore unidirectional and relations of power remain as they were: the teacher sets the agenda and direction for emancipation.

One example of such problems is illustrated as early as 1981, in Judith Williamson's article entitled, "How Does Girl Number Twenty Understand Ideology?" The original Girl Number Twenty was Sissy Jupe in Charles Dickens' novel *Hard Times*. Asked to define a horse by the tyrannical rationalist, Mr. Gradgrind, Sissy Jupe, who knows horses from her experience, remains silent. The definition was then given by a boy called Bitzer, who parroted off an encyclopedic definition of the horse:

> **Vignette:**
>
> [*Mr. Gradgrind:*] "Give me your definition of a horse."
>
> [...]
>
> (Sissy Jupe thrown into the greatest alarm by this demand.)
>
> "Girl number twenty unable to define a horse!" said Mr. Gradgrind, for the general behoof of all the little pitchers. "Girl number twenty possessed of no facts, in reference to one of the commonest of animals! Some boy's definition of a horse. Bitzer, yours."
>
> [...]
>
> "Bitzer," said Thomas Gradgrind. "Your definition of a horse." "Quadruped. Graminivorous. Forty teeth, namely twenty-four grinders, four eye-teeth, and twelve incisive. Sheds coat in the spring; in marshy countries, sheds hoofs, too. Hoofs hard, but requiring to be shod with iron. Age known by marks in mouth." Thus (and much more) Bitzer.
>
> "Now girl number twenty," said Mr. Gradgrind. "You know what a horse is." (Dickens, 1997, Chapter II, para. 16–22)

Williamson (1981) did not write about the definition of a horse and Sissy Jupe, but her starting point is a classroom experience, where three girls in a classroom sit in embarrassed silence when Williamson asked them to consider the "representation of women in the media: myth, convention and code." Williamson was deeply troubled by this unintended outcome of her attempt to empower the girls through an analysis of the types of popular culture, which she believed holds them in thrall to patriarchy through the workings of ideology. Her own "Girl Number Twenty" was Astrid. Williamson wrote the following:

> My worst problem is Astrid. She looks, actually, very much like the blonde heroines of the comic strip, her self-image is clearly bound to the things we seem to be attacking. She sits at the front of the class and says, literally, nothing. She may file her nails, or just stare: I'm really worried about her. I offer to teach her with the other girls, but she doesn't particularly want to. (p. 82)

Williamson added the following in the footnote: "Astrid is a good example of someone who could *not* learn about sexism *analytically....* " Williamson tried to empower students through the methods of progressive pedagogy, but what was missing was the knowledge of students' own life experiences and problems. The method of progressive pedagogy was still a *banking system*, if we use Paulo Freire's term.

Jacques Rancière (1991) tried to develop a radical learning theory that he called an emancipatory method. The key aspect of this method is radical equality between teacher and students, which for Rancière are both ignorant subjects. The relation between these is based on the capacity of speech and communication. Rancière suggested that an emancipatory method starts from the assumption that all students already speak and are speakers. "[Human] child is first of all a speaking being," he wrote (p. 11).

When Rancière wrote of the significance of voice, he often referred to the *secessio plebis* (literally, withdrawal of the commoners/plebeians) on the Aventine Hill in Rome (494 BCE) as the first scene or narration of the voice of the voiceless. It is the question of who has the right to speak and be recognized and heard as persons. In this story, plebeians gathered on the Aventine Hill and demanded a treaty with the patricians. Rancière (2004) wrote that

> the patricians responded that this was impossible, because to make a treaty meant giving one's word: since the plebeians did not have human speech, they could not give what they did not have. They possessed only a "sort of bellowing which was a sign of need and not a manifestation of intelligence." In order to understand what the plebeians said, then, it had first to be admitted that they spoke. And this required a novel perceptual universe, one where—contrary to all perceptible evidence—those who worked for a living had affairs in common with free men and a voice to designate and argue these common affairs. (p. 5)

What the plebeians did was give themselves a name and link their bodies to the practice of speech and language. They had the opportunity to tell their stories, life experiences, and life conditions. If we think of voice as a process, as some kind of action, it means the process of giving an account of one's life and its conditions (Couldry, 2010, p. 7). Couldry (2010, p. 7) wrote that such an account means telling a story, providing a narrative. This kind of narrative is a basic feature of human action.

> What we do—beyond a basic description of how our limbs move in space—already comes embedded in narrative, our own and that of others. This is why to deny value to anoth-

er's capacity for narrative—to deny her potential for voice—is to deny a basic dimension of human life. (Couldry, 2010, p. 7)

For Couldry, having a voice is not enough, but people have to know that their voice matters. In media education, an important objective is that students have a voice that matters, that they are part of school democracy and community. And that is how empowerment starts.

In his book, *The Ignorant Schoolmaster*, Rancière (1991) wrote about the 19th-century teacher Joseph Jacotot, who had an interesting pedagogical experience. Jacotot was a soldier, teacher, and politician during the French Revolution, who had to escape to the Netherlands, where he was employed by the University of Leuven to teach. Unfortunately, Jacotot spoke French while his students spoke Dutch and Flemish. Jacotot got an idea to offer students the bilingual *Télémaque* by Fénelon for reading. He asked students to read portions of the book with the help of Flemish, so that they could speak about it in French. After all, students had to write what they thought about the text in French. What Jacotot learned was that students wrote as well as many French-speaking natives. Jacotot understood that students are able to learn by themselves and, what is more revolutionary, that the teacher does not need to know everything or more than the students: the teacher can teach what he or she does not know.

For Ranciére, the story does not open a new pedagogical theory but is a sign of a radical equality. This equality is not a goal of all education but the starting point, axiom, of all education: "Everyone is of equal intelligence" (Rancière, 1991, p. 109). Students learn when they are treated as spoken citizens and when other people speak to them as equals. In a way, equality is the best teaching method and language is the only tool for teaching. Students learn when they want to learn and have room for learning. The same basic idea is crucial for media education. It is a place for students to use their voice and make meaning by working with the language of media.

New Culture of Learning

Thomas and Brown (2011) wrote about another story, very similar to Jacotot's. In the spring of 2004, Douglas Thomas taught a course entitled "Massively Multiplayer Online Games and the University of Southern California." The course started with traditional lectures and readings and the teacher let students also share experiences of the game *Star War Galaxies*. By the third week, students interrupted the teaching:

"Professor, I know we need to do the lecture and discussion, but we were all in the game last night and found the perfect example for class today, so can we please start with it and then do the lecture after?" Doug agreed, though he couldn't help feeling that his

> class was now slipping away from him. In one sense that was true. But he had missed a far more important message: Students were getting together, outside of the class, and discussing how they might make the class itself better. (Thomas & Brown, 2011, Chapter 1, "Teaching in a Galaxy Far, Far Away," para. 4)

Thomas first thought that he had failed as a teacher and could no longer control the classroom. But at the end of the course, when he read students' final exams, he noticed that all their papers were filled with examples of the students' gaming, were highly textured with references, and more importantly, they featured the students referring to each other. The students had formed their own learning community and space: "And they had done it by themselves, for themselves" (Chapter 1, "Teaching in a Galaxy Far, Far Away," para. 9).

When students are highly motivated, they find their way to learn and seek information. Sometimes it takes place without teachers, without books, and without classrooms. Thomas and Brown (2011) called teaching all around us "the new culture of learning." They wrote that, "learning should be viewed in terms of an environment" (Chapter 2, "Learning Environments," para. 1.) Learning is spatial, needs a rich space and environment of information and people, students, and also teachers, for a collective to emerge, where "all coexist and shape each other in mutually reinforcing ways" (Chapter 2, "Learning Environments", para. 1). Students learn in spaces *within* the world and not so much about the world; they learn *how* to get information rather than *what* of information, but also information about the world has to be learnt.

Learning Together in Both Sites of the School

We have seen that teaching and learning settings in the Maple School vary from teacher-driven situations to more dialogical learning. Media and ICT learning makes different learning situations possible and students brought their voices to the school by making their own media presentations both inside and outside of school. But teachers were needed as part of mutual learning space, not so much as masters but enablers of learning. I, myself, observed many times when teachers and students tried to solve problems together. Here is one vignette:

Three students and a teacher try to find a video file from the computer. Last time they had some problems with the computer:

> Tea is on the computer, Dolores on the right, and Saku on the left side:
>
> *Teacher:* What happened last time, didn't it accept your...?
>
> *Tea:* It didn't open at all. It didn't accept my account at all.
>
> *Teacher:* Okay, let's see...

Saku: What was the problem?

Teacher: They couldn't open the computer last time, or was it the file?

Tea: We didn't get anything open.

Teacher: And if it is not ready remember to save your file in any case.

Saku: Is it here in this computer?

Teacher: Yes.

They succeeded to open the computer. Tea tries to write her account but the computer makes a "beep" when she uses the keyboard.

Tea: Is there something on?

Saku: Click to the field.

Tea: Now I'm in.

Teacher: Sometimes that caps lock or something else is on and problems will arise because of that. But this didn't show it with any light indicator. I think this is okay.

Tea is not used to using a PC and does not know where to find a file.

Tea: Help. I want out of here.

Teacher: No, no, stay there.

Tea: I can't use this.

Saku: There, on the top right, file, open a project.

Tea: Where is it?

Saku: C or D disk.

Teacher: Take the cursor...

Tea: Wasn't it here?

Teacher: Probably you have put it there.

Saku: You can use Mac but not XP?

Tea: That's right.... I can't find the file.

Teacher: What name did you use when you saved it?

Dolores: If it is there?

Tea: It wasn't.

Tea: Marjo saved it.

Saku: Marjo should better to be here.

Tea: She didn't want to. (From my notebook)

What is interesting in this vignette is the fact that the computer is being operated by a girl who is most incapable of using it because she has used a totally different operating system at home. At the same time, the teacher is not somebody who "knows all" and just gives orders. Rather, they try to solve the problem together by the logic of trial and error. In fact, the teacher helps Tea solve the problem by not letting her give up and go. As we see, using ICT in an official school space requires group work, where the teacher is not an expert. There may be some other experts who help, like in this case Saku, who helped Tea at the start.

In the same way, in the movie course, students sometimes asked Janne for help, who seemed to be a kind of "technical guru" for the class. Janne sometimes helped others with editing software but at the same time asked for help from the teacher, who said that she had used Final Cut Pro a long time ago and "had to dig somewhere from the bottom of memory" on how to use the software. Usually negotiation and the method of trial and error helped solve issues, but of course, this is not an ideal way to solve problems in school settings.

Working with ICT and digital technology at school opens traditional teacher-oriented teaching to more peer-to-peer learning. This is even more common in unofficial school space and outside of school.

Collegial Pedagogy

I have been discussing a lot of different changes, especially regarding teaching, learning, and school spaces, from content and subject-based "what" learning to context and participation-based "how" learning, learning to be and understanding the context of meaning. Peer-based learning is one part of this change, but we should not dismiss the meaning of teachers who give safe space for learning and are facilitators who help students find new perspectives and scaffolds of learning.

Soep and Chávez (2010) described the relationship between learners and teachers in terms of *collegial pedagogy,* which brings young people and adults into productive collaborative relationships in the communities of practice. Soep and Chávez's context is Youth Radio, where adults act as producers and provide guidance for beginners in media production. The important point is that adults and youth need each other to be successful:

> The adult producer cannot create the story without young people to identify topics worth exploring, to find and interview characters, and to experiment with novel modes

> of expression and ways of using words, scene, and sound. At the same time, young people cannot create a story without adults to provide access to resources, equipment, broadcast outlets, and institutional recognition and share the skills and habits developed through years of experience as media professionals. (Soep & Chávez, 2010, p. 55)

While teachers' roles in a school are not exactly the same as adult producers, they provide both opportunities and perspectives for students that are not possible elsewhere. For example, in the Maple School, the visual arts teacher used a Media School that was not accessible without the teacher and school context. I observed that some girls who were not very interested in media production in the first place found their "inner pro-ams" during the hands-on working portion of the project and started to lead production.

One example is 15-year-old Peppi, who had problems with textual school subjects and was under teachers' supervision. In the movie course, she was always ready to do things, and as her visual arts teacher said, she was a "contrast in that she doesn't have enough strength in other school subjects."

Soep and Chávez (2010, p. 54) emphasized that collegial pedagogy draws from a framework of (a) community of practice, (b) critical pedagogy, and (c) positive youth development. Community of practice is drawn from the social theory of learning by Lave and Wenger (1991), who emphasized the process of participation in learning settings. Such participation shapes what we do, what we are, and how we interpret what we do and are. Communities of practice are integral parts of everyday life but mostly "so informal and so pervasive that they rarely come into explicit focus" (Wenger, 2008, p. 7):

> Students go to school and, as they come together to deal in their own fashion with the agenda of imposing institution and the unsettling mysteries of youth, communities of practice sprout everywhere—in the classroom as well as on the playground, officially or in the cracks. And in spite of curriculum, discipline, and exhortation, the learning that is most personally transformative turns out to be the learning that involves membership in these communities of practice. (p. 6)

I have given examples of young people who participate in these kinds of communities of practice: they remix and create digital stories in different ways in varieties of communities on the Internet and in relation to the school and school community. Young people want to be recognized, valued, and heard as persons. Their creative production is mostly a social activity and connectivity with like-minded others.

Soep and Chávez (2010, p. 54) emphasized that learning is something that communities create, not something that teachers transmit. Rather, students and teachers work together. What teachers can especially bring to the learning settings are themselves as *public and border intellectuals.* This term is used in the context of critical pedagogy and by Henry Giroux. He referred to teachers as intellectuals who research and think with the students and help them understand cultural and social relations and states of affairs and how and why our relations to the world are as they are. Public and border intellectuals "travel within and across communities of difference working in collaboration with diverse groups and occupying many sites of resistance" (Giroux, 1996, p. 54). They create new spaces where students can discuss, change ideas, learn from each other, and create resistance against social injustice and economic equality. According to Giroux, a school and education can be spaces in which the dominant culture is encountered and challenged by students and teachers as border intellectuals:

> Elements of resistance now become the focal point for the construction of different sets of lived experiences, experiences in which students can find a voice and maintain and extend the positive dimensions of their cultures and histories. (Giroux, 2001, p. 112)

Soep and Chávez (2010, p. 56) argued that audience can especially radicalize youth media pedagogy. This is especially in question when young people are really able to use their voice and influence public debate and even social policy. Collegial pedagogy draws on a pedagogy of positive youth development, "which plays to young people's strengths and counters the tendency to criminalize and pathologize young people" (p. 56). At the same time, it tries to avoid fetishizing the voice of the youth and celebrating the lives of young people as something especially authentic. There is always a tendency that the experiences of the youth are sensationalized and marginalized. Therefore, educators are important, as adults or peers, who share the production process.

School subjects like history, social science, visual arts, and mother tongue are places to enhance students' critical thinking and understanding of media. Instead, Peer-to-peer productions outside the school are mostly "learning to be." "Learning to be" refers to social participation in peer-based learning. Brown and Adler (2008) emphasized learning especially on digital networks as social in nature: understanding of content is socially constructed in networks and conversations with others. The focus here is more on *how* people are learning than on *what* they are learning. Emphasis is on human interactions around the content. According to Brown and Adler, perhaps the more important aspect of social learning is "'learning to be' a full participant on the field" (p. 19). This requires participants to acquire the practices and norms of established practitioners or acculturate into a community of practice. Learning is based on an apprenticeship where a learner first watches and takes notes before simple and demanding tasks. Examples of this kind of learning usually come from open source communities, the Wikipedia community (Suoranta & Vadén, 2010), or fan fiction communities (Black, 2008).

Learning to be is a valuable skill, but usually in the case of media education, it does not touch the core aspect: learning about the media. Learning about the media has its own goals and agendas defined by school administrations and teachers. Learning about the media is school-based "learning about the subject." However, these two aspects of learning are not necessarily in contrast; they often support each other. Young people bring their media practices, interest areas, and cultural knowledge to the school, but teachers have a responsibility to guide questions toward a more critical direction and to be the more experienced individual in a student-teacher relationship. This kind of pedagogy balances peer-to-peer learning and youth-teacher collaboration (cf. Soep & Chávez, 2010) and makes a bridge between official and unofficial school spaces. Soep and Chávez (2010) wrote that a key goal of collegial pedagogy is that "young people develop the technical, creative, and intellectual capacities that enable them once again to step away from heavy adult involvement" (p. 52). In good circumstances, collegial pedagogy can open a positive circle where students bring "vernacular literacies" and their knowing *how* to the school via the unofficial school spaces, and teachers are able to utilize this knowledge and skills in official school spaces. This may lead to a process whereby students learn new literacy practices and critical thinking in official school spaces in order to transfer this again to their independent out-of-school-based media productions.

Assessing Media Literacies

After all, we have to ask this: Do we know that teaching media education can enhance media literacy of the youth or do they learn all that is necessary out of school? Young people, as digital natives, are often considered to have better digital and media literacy skills than previous generations. In his classic book, *Growing Up Digital,* Don Tapscott (1999) argued that children have become authorities on technology for the first time in human history. He saw a huge generation gap between earlier and the so-called *Net* generation in the same way as Marc Prensky (2001). However, there is little evidence that young people have superior skills compared to adults.

In her study about visual literacies, Eva Brumberger (2011) argued that the youth do not possess a high degree of visual literacy: "Their ability to respond critically to visual material appears rather weak" (p. 44). Even students' self-assessment of their skills with technologies for producing visual communication was limited. In addition, research into young people's Internet information skills shows that they lack evaluative and strategic skills (Sonck, Kuiper, & de Haan, 2012).

In the Maple School, I asked students to assess their technological and literacy skills by grading themselves between numbers 4 and 10, which are commonly used in their school (10 = best, 4 = fail) (see Table 6.1). According to students' own accounts of their media skills they seemed to be quite confident with their abilities in every area. They even trusted their skills in the area of office-based programs like word processing and spreadsheet computation. Self-assessing is not, of course, as valid as tests for evaluating media and digital literacy but this result gives still different picture of media skills of young people than, for example,

Table 6.1: Students' Self-Assessment of Their Media Skills ($N = 305$)

Skills	Grade
Searching for information for schoolwork	8.6
Assessing the reliability of information on the Internet	8.0
Altering images	8.0
Uploading images or videos to the Internet	8.0
Designing media presentations	7.6
Computing with a spreadsheet computation	7.0
Using a word processor	8.4
Doing remixes and mashups	5.3
Keeping personal information and passwords secret on the Internet	9.2
Editing Internet pages and profile sites	8.0
Modifying and customizing games	5.3

Brumberger's study. But this tells more how students rely on their own skills and not at all, for example, about their analytical and critical skills.

The highest rating students gave themselves in the Maple School was in the area of Internet safety. In the EU Kids Online study, Finnish children and young people claimed critical and safety skills more than children in any of the other 24 European countries; they claimed 5.8 of 8 skills, when the average in Europe was 4.2 (Livingstone et al., 2011, p. 28).

The poorest ratings that students gave themselves was in the area of creative production, like remixing and modifying games; both scored 5.3. Self-assessment, of course, is not an easy task to do, but we may also say that people usually tend to overestimate their technological skills (Brumberger, 2011, p. 44). From this perspective, the students' creative skills may be even weaker than the survey suggests.

In the Maple School, the visual arts teacher suggested that students usually draw more attention to the frivolous story and not to the visual and symbolic meaning of the images. When she talked about an assignment where the task was to create stories from still images, she mentioned that students could not create new "multifaceted and multilayered stories"; on the contrary, they played like "this could do this, and then would have a baby and get married."

Interviewer: Like a play?

Teacher: Yes, that came to my mind, that "we just play with these images."

Interviewer: Like with paper dolls?

Teacher: Exactly. (Interview with the visual arts teacher, female)

According to the teacher students seem to have adopted some kind of playful attitude toward images and not the critical and thoughtful attitude that she wanted them to have. She mentioned that some of her other classes were different, and the students there had more discussions and thoughts.

Obviously, students had different kinds of knowledge and thoughts about media and the Internet. I observed some eighth-grade girls discussing positive and negative aspects of social networking sites like Facebook. They listed risks of using Facebook, which included getting a virus, forgetting schoolwork, spending too much time on Facebook, risking misuse of their personal information, getting hooked on Facebook, being too young to use Facebook, and being stalked. These were critical issues that they clearly considered as risky when using social networking sites. Finnish youngsters are especially careful about their personal information on social networking sites. The EU Kids Online survey indicated that Finnish children and teenagers had experienced data misuse least often among European countries. Only 5% of young people reported having experienced data misuse, while the international country average was 10% (Kupiainen, Suoninen, & Nikunen, 2011). Still, only 74% 11- to 16-year-old Finns reported that they can change privacy settings on the social networking profiles (Uusitalo, Vehmas, & Kupiainen, 2011, p. 123), but the international country average was even lower, at only 56% (Livingstone et al., 2011a, p. 27).

What I can say is that school matters when it comes to media skills and literacy. Buckingham (2010) argued that schools have a big role in addressing the participation gap and critical literacy skills. This is much more than help in accessing the Internet and equipment: it is also about cultural competencies, social skills and especially knowledge of the media; "about how media work, about the media industries, about the history of media, about the uses and effects of media within society" (p. 298). In the Maple School, visual arts and mother tongue teachers tried to enhance especially students' critical literacy skills.

We have evidence that teaching the cross-curricular theme of media skills and communication in a school enhances students' media literacy skills. The Finnish National Board of Education carried out an extensive study assessing the goals of cross-curricular subjects in ninth grade. The survey was conducted in 2010 with over 8,000 students in 113 schools in Finland. The assessment of the goals of media skills and communication was implemented with 1,206 Grade 9 students. The survey indicated that students who participated in cross-curricular studies of media skills and communication had better results in media knowledge (Kotilainen & Kupiainen, 2012). Eighty-two percent of students who had participated in cross-curricular teaching during school lessons solved knowledge questions compared to those 78% of who had not participated in the teaching and solved the questions. The difference is not huge but statistically significant. There was also the same kind of difference between students who had studied media themes at home and outside of school and those who had not.

Media Education in Heterotopia

If we think of assessment from the perspective of new literacy studies, we encounter special difficulties, especially because new media practices are not necessarily end-product oriented and they are related to social learning comprising processes characteristic of multimodal texts (Burke & Hammett, 2009, p. 7). Changing literacies, texts, and schools requires changing assessments as well (p. 7). Burke and Hammet suggested (2009) that we need to move from traditionally assessing only "snapshots of particular points of learning in particular populations at particular times" to an understanding of "youth's practices and development of texts," how they participate and communicate using literacy practices (p. 7). This has been one task of this book. Trying to assess students' media literacies in the traditional way is out of this task.

Even though differences in learning outcomes among Finnish schools are relatively small, there are differences especially in the area of media education, which does not have the status of an official school subject. In this study, I have also reported differences between teachers and school subjects. Media education is especially a focus in mother tongue, visual arts, history, social science, and joint ICT classes, but in different ways. Assignments in mother tongue classes were based mostly on textbooks, and pedagogical practices varied from simulation and critical textual analysis to media production. Visual arts teachers concentrated more on media production but used analysis as a means to understand storytelling. In history and social science, media production was not so much learning about the media but with the media, using media production for school assignments and learning about the school subjects. Joint ICT classes were more about learning different computer programs and using them for students' own purposes. One common theme for all classes in the 2009–2010 academic year was advertisements as an object of analysis and media production in order to enhance students' understanding of the impacts of advertisements in various forms.

If we think of media education from the perspective of heterotopia, we can see its heterogeneity and ambiguity, as a space in terms of the 3-Cs model of media literacy: cultural, creative, and critical. Young people as well as adults and teachers are living in consumer culture and using media devices and contents for different purposes. One main task of digital devices is to create new consumer culture where media contents like games, movies, magazines, Internet communities, television programs, music, and so forth, are produced by traditional companies and experienced and consumed by different audiences. A culture provides a context for identity building, consumption habits, discussions, resources for remixing and media productions, and criticism. Creative and critical aspects of media literacy are tied to this cultural context. Critical questions and analysis take place in everyday culture. It should not be pure academic discourse without any bonds to the lifeworld of the youth.

With my colleague Sara Sintonen (Kupiainen & Sintonen, 2010), we have offered a design model for media education, a circle that starts from "culture," where students identify cultural modes of media texts and culture in themselves and in their identity play and taste. Media education gives the possibility to *design* and *re-sign* cultural meanings, remix texts and contexts in order to make new meanings in new contexts. From the basis of this study, I can say that teachers play an important role as colleagues who step up to the heterotopic media practices and value the processes and learning that come from outside the school. In the school context, new complex spaces and spatial practices are needed in order to bridge different cultures and learning processes.

As a circle, the design model moves between culture(s), creation, and critical activities, and is a kind of hermeneutic cycle that characterizes understanding. Hermeneutic cycle refers here to the idea that individual media phenomena are understandable only in relation to the broader cultural context and vice versa. Therefore, it is a circle. Hermeneutics is a traditional interpretation theory of texts, an "art of understanding," in a way where interpretation must proceed by framing the piece of text in terms of the overall organization of the text. We do not understand hermeneutics so much as a methodological approach but a philosophical term that refers to the human understanding. From this perspective, like Martin Heidegger (1988) wrote, "any interpretation which is to contribute understanding, must already have understood what is to be interpreted" (p. 194). When we are interpreting something, we have some fore-structures, some basic knowledge and ways of understanding at the starting point. Interpretation does not start from null. Also, we do not have to limit ourselves to standard assumptions, but we can check our beliefs and understanding against other beliefs and understanding. This is the way to get new information and *re-sign* semiotic resources for our own purposes as people do when they, for example, remix popular texts. The design model encourages the process of making and remaking, reshaping and remixing, designing media stories and meanings so that people can feel that their voice has a meaning and that they have recognition and are respected as contributors.

This study has suggested that school is changing little by little via the media practices that children and young people bring into it when they open new spaces for their creative practices. But students need teachers' help in order to enhance their skills and to understand the media and surrounding culture and society.

Dominant media culture serves many possibilities for diverse media consumption and participatory activities. But participatory culture is not self-evident; it also needs critical reflection.

Finnish schools have focused on subject-based learning and knowledge building that is partly in the background of good PISA results. As teachers in the Maple School have pointed out many times, textbooks are heavier and there is more and more to teach within each school subject. Time for creative practices, critical analysis, and media productions is a limited resource. At the same time, in the age of financial crisis, educational sectors suffer from budget cuts. While writing this, I heard that the aforementioned Media School, which offered resources and pedagogic help for media education in a city area, might be closed. Because media is such an inseparable part of the everyday lives of the youth, it may be that media education is not deemed necessary from the perspective of school policy. Despite researchers' comments about the importance of 21st-century skills and needs for changing pedagogical paradigms and learning practices at schools, the business-as-usual thinking seems to have a dominant role in school administration: knowledge and learning *what* is more important than learning *how* and all practices at school are spatialized within existing time-space structures. The Maple School had some advantages and students and teachers some emphasis on creative practices and other-space production. Their media literacy practices were productive of the space and also made room for new learning practices. This was the way to create heterotopic spaces for learning.

REFERENCES

Ahonen, S., & Virta, A. (1999). Toward a dynamic view of society: Civic education in Finland. In J. Torney-Purta, J. Schwille, & J.-A. Amadeo (Eds.), *Civic education across countries: Twenty-four national case studies from the IEA Civic Education project* (pp. 229–156). Amsterdam, the Netherlands: The International Association for the Evaluation of Educational Achievement.

Appadurai, A., & Breckenridge, C. A. (1995). Public modernity in India. In C. A. Breckenridge (Ed.), *Consuming modernity: Public culture in a South Asian world* (pp. 1–20). Minneapolis: University of Minnesota Press.

Asthana, S. (2011). Youth media imaginaries in the Arab world: A narrative and discourse analysis. In. J. Fisherkeller (Ed.), *International perspectives on youth media. Cultures of production and education* (pp. 50–67). New York, NY: Peter Lang.

Barton, D. (2007). *Literacy. An introduction to the ecology of written language* [Kindle iPad version]. Oxford, UK: Blackwell. Retrieved from http://www.amazon.com

Beck, U. (1992). *Risk society: Towards a new modernity* (M. Ritter, Trans.). Newbury Park, CA: Sage.

Bennett, S., Maton, K., & Kervin, L. (2008). The "digital natives" debate: A critical review of the evidence. *British Journal of Educational Technology, 39*(5), 775–786.

Bhabha, H. (1994). *The location of culture.* New York, NY: Routledge.

Bigum, C., & Rowan, L. (2008). Landscaping on shifting ground: Teacher education in a digitally transforming world. *Asia-Pacific Journal of Teacher Education, 36*(3), 245–255.

Black, R. (2008). *Adolescents and online fanfiction.* New York, NY: Peter Lang.

Black, R. W. (2009). Just don't call them cartoons. The new literacy space of anime, manga, and fanfiction. In J. Coiro, M. Knobel, C. Lankshear, & D. Leu (Eds.), *Handbook of research on new literacies* (pp. 583–610). New York, NY: Routledge.

Borgmann, A. (1984). *Technology and the character of contemporary life: A philosophical inquiry.* Chicago, IL: University Of Chicago Press.

boyd, d. (2008). Why youth [love] social network sites: The role of networked publics in teenage social life. In D. Buckingham (Ed.), *Youth, identity, and digital life* (pp. 119–142). Cambridge, MA: The MIT Press.

boyd, d., & Ellison, N. (2007). Social network sites: Definition, history, and scholarship. *Journal of Computer-Mediated Communication, 13*(4), 210–230.

Brown, J. S., & Adler, R. P. (2008, January/February). Minds on fire: Open education, the long tail, and learning 2.0. *Educause Review, 43*(1), 16–32. Retrieved from http://www.johnseelybrown.com/mindsonfire.pdf

Bruijn, M., Nymanjoh, F., & Brinkman, I. (Eds.). (2009). *Mobile phones: The new talking drums of everyday Africa.* Cameroon, Africa: Langaa Research and Publishing Common Initiative Group.

Brumberger, E. (2011). Visual literacy and digital native: An examination of the millenial learner. *Journal of Visual Literacy, 30*(1), 19–46.

Bruns, A. (2008). *Blogs, Wikipedia, Second Life, and beyond.* New York, NY: Peter Lang.

Buchanan. I. (1993). Extraordinary spaces in ordinary places: De Certeau and the space of postcolonialism. *Journal of the South Pacific Association for Commonwealth Literature and Language Studies, 36,* X. Retrieved from http://wwwmcc.murdoch.edu.au/ReadingRoom/litserv/SPAN/36/Jabba.html

Buckingham, D. (2000). *After the death of childhood. Growing up in the age of electronic media.* Cambridge, UK: Polity Press.

Buckingham, D. (2003a). Media education and the end of critical consumer. *Harvard Educational Review, 73*(3), 303–327.

Buckingham, D. (2003b). *Media education. Literacy, learning and contemporary culture.* Cambridge, UK: Polity.

Buckingham, D. (2010). Do we really need media education 2.0? Teaching media in the age of participatory culture. In K. Drotner & K. Schroder (Eds.), *Digital content creation. Perceptions, practices & perspectives* (pp. 287–304). New York, NY: Peter Lang.

Buckingham, D. (2011). Foreword. In M. Thomas (Ed.), *Deconstructing digital natives. Young people, technology and the new literacies* (pp. ix–xi). New York, NY: Routledge.

Burke, A., & Hammett, R. F. (2009). *Assessing new literacies.* New York, NY: Peter Lang.

Burn, A., & Durran, J. (2007). *Media literacy in schools: Practice, production and progression.* London, UK: Paul Chapman.

Busse, K. (2008). My life is a WIP on my LJ: Slashing the slasher and the reality of celebrity and Internet performances. In K. Hellekson & K. Busse (Eds.), *Fan fiction and fan communities in the age of the Internet* (pp. 207–224). Jefferson, NC: McFarland & Company. Retrieved from http://www.amazon.com

Carpentier, N. (2010). The reception of the "producers'" films on a participatory website. Ordinary young people and the politics of banality. In T. Olsson & P. Dahlgren (Eds.), *Young people, ICTs and democracy* (pp. 51–68). Gothenburg, Sweden: University of Gothenburg, Nordicom.

Carpentier, N. (2011). *Media and participation. A site of ideological-democratic struggle* [Kindle iPad version]. Chicago, IL: Intellect. Retrieved from http://www.amazon.com

Carpentier, N., & de Cleen, B. (2008). Introduction: Blurring participations and convergences. In N. Carpentier & B. de Cleen (Eds.), *Participation and media production: Critical reflections on content creation* (pp. 1–12). Newcastle, UK: Cambridge Scholars.

Casey, E. S. (2009). *Getting back into place. Toward a renewed understanding of the world* (2nd ed.). Bloomington: Indiana University Press.

Castells, M. (1996). *The rise of the network society*. Cambridge, UK: Blackwell.

Chittenden, T. (2010). Digital dressing up: Modelling female teen identity in the discursive spaces of the fashion blogosphere. *Journal of Youth Studies, 13*(4), 505–520.

Christensen, O., & Tufte, B. (2010). Media education: Between theory and practice. In S. Kotilainen & S. Arnolds-Granlund (Eds.), *Media literacy education: Nordic perspectives* (pp. 109–120). Gothenburg, Sweden: University of Gothenburg, Nordicom.

Couldry, N. (2008). Media and the problem of voice. In N. Carpentier & B. de Cleen (Eds.), *Participation and media production: Critical reflections on content creation* (pp. 15–26). Newcastle, UK: Cambridge Scholars.

Couldry, N. (2010). *Why voice matters. Culture and politics after neoliberalism*. Thousand Oaks, CA: Sage.

Currie, C., Gabhainn, S. N., Godeau, E., Roberts, C., Smith, R., & Currie, D. (Eds.). (2008). *Inequalities in young people's health: HBSC international report from the 2005/2006 Survey. Health policy for children and adolescents, No. 5.* Copenhagen, Denmark: World Health Organization. Retrieved from http://www.euro.who.int/__data/assets/pdf_file/0005/53852/E91416.pdf

Currie, C., Roberts, C., Morgan, A., Smith, R., Settertobulte, W., Samdal, O., & Barnekow Rasmussen, V. (Eds.). (2004). *Young people's health in context: Inter- national report from the HBSC 2001/02 survey. Health policy for children and adolescents, No. 4.* Copenhagen, Denmark: World Health Organization. Retrieved from http://www.hbsc.org/publications/reports.html

De Certeau, M. (1984). *The practice of everyday life* (S. Rendall, Trans.). Berkeley: University of California Press.

Dean, J. (2010). *Blog theory. Feedback and capture in the circuits of drive*. Cambridge, UK: Polity Press.

Derecho, A. (2008). Archontic literature. A definition, a history, and several theories of fan fiction. In K. Hellekson & K. Busse (Eds.), *Fan fiction and fan communities in the age of the Internet* [Kindle iPad version]. Jefferson, NC: McFarland & Company. Retrieved from http://www.amazon.com

Dickens, C. (1997). *Hard times* [Kindle iPad version]. Champaign, IL: The Project Gutenberg. Retrieved from http://www.amazon.com

Dover, C. (2007, May). Everyday talk: Investigating media consumption and identity amongst school children. *Particip@tions*, 4(1). Retrieved from http://www.participations.org/Volume%204/Issue%201/4_01_dover.htm

Eisenhart, M. (2001). Educational ethnography past, present, and future: Ideas to think with. *Educational Researcher, 30*(8), 16–27.

Erstad, O. (2008). Trajectories of remixing—digital literacies, media production and schooling. In C. Lankshear & M. Knobel (Eds.), *Digital literacies: Concepts, policies and practices* (pp. 177–202). New York, NY: Peter Lang.

European Commission. (2007). *Study on the current trends and approaches to media literacy in Europe*. Retrieved from http://ec.europa.eu/culture/media/literacy/docs/studies/study.pdf

Finnish National Board of Education (2011.) *Lapset ja nuoret mediaosallistujina – osallistumista tukevan mediakasvatuksen toimintaohjelma* [Children and young people as participant on media – Action plan of participation supporting media education] (Reports 2011:28). Helsinki:

Finland: Helsinki, Finnish National Board of Education. Retrieved from http://www.oph.fi/download/139651_Lapset_ja_nuoret_mediaosallistujina.pdf

Fenton, N. (2010). Re-imagining democracy, new media, young people, participation and politics. In T. Olsson & P. Dahlgren (Eds.), *Young people, ICTs and democracy* (pp. 19–34). Gothenburg, Sweden: University of Gothenburg, Nordicom.

Fisherkeller, J. (Ed.). (2011). *International perspectives on youth media. Cultures of production and education.* New York, NY: Peter Lang.

Fiske, J. (1992). The cultural economy of fandom. In L. A. Lewis (Ed.), *The adoring audience* (pp. 30–49). New York, NY: Routledge.

Foucault, M. (1967). Of other spaces, heterotopia (J. Miskoviec, Trans.). Retrieved from http://foucault.info/documents/heteroTopia/foucault.heteroTopia.en.html

Foucault, M. (1992). *The order of things: An archaeology of the human sciences.* New York, NY: Routledge.

Foucault, M. (2008). *This is not a pipe* (J. Harkness, Trans.). Berkeley: University of California Press.

Garrison, M. J., & Bromley, H. (2004). Social contexts, defensive pedagogies, and the (mis)uses of educational technology. *Educational Policy, 18*(4), 589–613.

Gauntlett, D. (2011). *Making is connecting. The social meaning of creativity, from DIY and knitting to YouTube and web 2.0.* Cambridge, UK: Polity Press.

Gee, J. P. (1991). What is literacy? In C. Mitchell & K. Weiler (Eds.), *Rewriting literacy: Culture and the discourse of the other* (pp. 3–11). Westport, CT: Bergin & Garvey.

Gee, J. P. (2004). *Situated language and learning: A critique of traditional schooling.* New York, NY: Routledge.

Gee, J. P. (2007). *What video games have to teach us about learning and literacy. Second edition: Revised and updated edition.* New York, NY: Palgrave Macmillan.

Gee, J. P. (2010). *New digital media and learning as an emerging area and "worked examples" as one way forward* [Kindle iPad version]. Retrieved from http://www.amazon.com

Gilster, P. (1997). *Digital literacies.* New York, NY: Wiley Computer.

Giroux, H. (1996). Is there a place for cultural studies in colleges of education? In H. Giroux, C. Lanhshear, P. McLaren, & M. Peters (Eds.), *Counternarratives, cultural studies and critical pedagogies in postmodern spaces* (pp. 40–58). New York, NY: Routledge.

Giroux, H. (2001). *Theory and resistance in education: Towards a pedagogy for the opposition* (Rev. ed.) [Kindle iPad version]. Westport, CT: Bergin & Garvey. Retrieved from http://www.amazon.com

Goffman, E. (1961). *Asylums: Essays on the social situation of mental patients and other inmates.* Garden City, NY: Anchor Books.

Gomez, M., Stone, J., & Hobbel, N. (2004). Textual tactics of identification. *Anthropology and Education Quarterly, 35*(4), 391–410.

Gordon, T., Holland, J., & Lahelma, E. (2000). *Making spaces: Citizenship and differences in schools.* New York, NY: St. Martin's Press.

Gordon, T., Holland, J., & Lahelma, E. (2001). Ethnographic research in educational settings. In P. Atkinson, A. Coffey, S. Delamont, J. Lofland, & L. Lofland (Eds.), *Handbook of ethnography* (pp. 188–203). Thousand Oaks, CA: Sage.

Gustavson, L. (2008). Influencing pedagogy through the creative practices of youth. In M. L. Hill & L. Vasudevan (Eds.), *Media, learning, and sites of possibility* (pp. 81–114). New York, NY: Peter Lang.

Hamilton, M. (2000, July). *Sustainable literacies and the ecology of lifelong learning.* Paper presented at the Supporting Lifelong Learning: A Global Colloquium Conference, London, UK. Retrieved from http://www.open.ac.uk/lifelong-learning/papers/393CCAC1-000B-67AA-0000015700000157_MaryHamilton-paper-noabstract.doc

Harkness, J. (2008). Translator's introduction. In M. Foucault, *This is not a pipe* (pp. 1–12). Berkeley: University of California Press.

Hartmann, M. (2010). Media literacy/competence, participation and youth. Conceptual reflections 2.0. In T. Olsson & P. Dahlgren (Eds.), *Young people, ICTs and democracy* (pp. 107–126). Gothenburg, Sweden: University of Gothenburg, Nordicom.

Haythornthwaite, C., & Wellman, B. (2002). The Internet in everyday life: An introduction. In C. Haythornthwaite & B. Wellman (Eds.), *The Internet in everyday life.* Oxford, UK: Blackwell Publishers.

Heath, S. B., & Street B. V. (2008). *Ethnography. Approaches to language and literacy research.* New York, NY: Teachers College Press.

Heidegger, M. (1988). *Being and time* (J. Macquarrie & E. Robinson, Trans.). Oxford, UK: Basil Blackwell.

Hetherington, K. (1997). *The badlands of modernity: Heterotopia and social ordering* [Kindle iPad version]. New York, NY: Routledge. Retrieved from http://www.amazon.com

Hine, C. M. (2000). *Virtual ethnography.* Thousand Oaks, CA: Sage.

Hobbs, R. (2004). Does media literacy work? An empirical study of learning how to analyze advertisements. *Advertising & Society Review, 5*(4), 1–28.

Hodas, S. (1993, September 14). Technology refusal and the organizational culture of schools. *Education Policy Analysis Archives, 1*(10). Retrieved from http://epaa.asu.edu/ojs/article/viewFile/687/809

Hodge, R., & Tripp, D. (1986). *Children and television. A semiotic approach.* Cambridge, UK: Polity Press.

Ilomäki, L., Taalas, P., & Lakkala, M. (2012). Learning environment and digital literacy. A mismatch or a possibility from Finnish teachers' and students' perspective. In P. P. Trifonas (Ed.), *Learning the virtual life. Public pedagogy in digital world* (pp. 63–78). New York, NY: Routledge.

Ito, M., Baumer, S., Bittanti, M., boyd, D., Cody, R., Herr-Stephenson, B.,...Tripp, L. (2010). *Hanging out, messing around, and geeking out: Kids living and learning with new media.* Cambridge, MA: The MIT Press.

Jenkins, H. (1992). *Textual poachers. Television fans & participatory culture* (Kindle iPad version). New York, NY: Routledge. Retrieved from http://www.amazon.com

Jenkins, H. (2006a). *Convergence culture: Where old and new media collide*. New York, NY: New York University Press.

Jenkins, H. (2006b). *Fans, bloggers and gamers: Exploring participatory culture*. New York, NY: New York University Press.

Jenkins, H., Clinton, K., Purushotma, R., Robison, A. J., & Weigel, M. (2006). *Confronting the challenges of participatory culture: Media education for the 21st century*. Chicago, IL: MacArthur Foundation. Retrieved from http://digitallearning.macfound.org/atf/cf/%7B7E45C7E0-A3E0-4B89-AC9C-E807E1B0AE4E%7D/JENKINS_WHITE_PAPER.PDF

Jones, R. (2005). Sites of engagement as sites of attention: Time, space and culture in electronic discourse. In S. Norris & R. H. Jones (Eds.), *Discourse in action. Introducing mediated discourse analysis* [Kindle iPad version]. New York, NY: Routledge. Retrieved from http://www.amazon.com

Jordan, S. (2003). Critical ethnography and the sociology of education. In C. A. Torres & A. Antikainen (Eds.), *The international handbook on the sociology of education: An international assessment of new research and theory* (pp. 82–100). Lanham, MD: Rowman & Littlefield Publishers.

Kairavuori, S., & Sintonen, S. (2012). Arts education: Instruments of expression and communication. In H. Nemi, A. Toom, & A. Kallioniemi (Eds.), *Miracle of education. The principles and practices of teaching and learning in Finnish schools* (pp. 209–223). Rotterdam, the Netherlands: Sense Publishers.

Kankaanranta, M., & Puhakka, E. (2008). *Kohti innovatiivista tietotekniikan opetuskäyttöä, Kansainvälisen Sites 2006-tutkimuksen tuloksia*. [Towards an innovative use of ICT in teaching. Results of international Sites research]. Jyväskylä: Finland: University of Jyväskylä, Finnish Institute for Educational Research. Retrieved from http://ktl.jyu.fi/img/portal/13816/SITES-julkaisu.pdf?cs=1228198530

Kant, I. (1900). *On education* (A. Churton, Trans.). Boston, MA: D. C. Heath and Company. Retrieved from http://files.libertyfund.org/files/356/0235_Bk.pdf

Karpovich, A. I. (2008). The audience as editor. The role of beta readers in online fan fiction communities. In K. Hellekson & K. Busse (Eds.), *Fan fiction and fan communities in the age of the internet* (pp. 171–188). Jefferson, NC: McFarland & Company. Retrieved from http://www.amazon.com

Katz, I. (2012, April 16). Battle for the Internet: Web freedom faces greatest threat ever, warns Google's Sergey Brin. Retrieved from http://www.guardian.co.uk/technology/2012/apr/15/web-freedom-threat-google-brin

Keen, A. (2007). *The cult of the amateur: How today's internet is killing our culture* [Kindle iPad version]. Garden City, NY: Doubleday. Retrieved from http://www.amazon.com

Kiili, C., Laurinen, L., & Marttunen, M. (2008). Students evaluating Internet sources: From versatile evaluators to uncritical readers. *Journal of the Educational Computing Research, 39*(1), 75–95.

Koskela, H. (2009). Webcams, TV shows and mobile phones: Empowering exhibitionism. *Surveillance & Society, 2*(2/3), 199–215.

Kotilainen, S., & Kupiainen, R. (2012). Viestintä- ja mediataidot [Communication and media skills]. In E. Niemi (Ed.), *Aihekokonaisuuksien tavoitteiden toteutumisen seuranta-arviointi 2010* [Assessment of the goals of cross-curricular themes 2010] (Education monitoring report 2012: 1) (pp. 107–124). Opetushallitus.

Kotilainen, S., & Suoranta, J. (2007). Mot en dialogisk mediepedagogikk [Towards a dialogical media education]. In S. Vettenranta (Ed.), *Mediadanning og mediepedagogikk. Fra digital begeistring til kritisk dommerkraft* [The media creation and media education. From digital enthusiasm to critical judgment] (pp. 104–120). Oslo, Norway: Gyldendal akademisk.

Kozinets, R. V. (2010). *Netnography: Doing ethnographic research online.* Thousand Oaks, CA: Sage.

Kupiainen, R. (2011). Young people, mobile phones and creative media practices at school. *Barn,* 3–4, 151–167.

Kupiainen, R. (in press-a). Young people's creative online practices in the context of a school community. *Cyberpsychology: Journal of Psychosocial Research on Cyberspace.*

Kupiainen, R. (in press-b). Dissolving the school space. Young people's media production in and outside of school. *Policy Futures in Education.*

Kupiainen, R., & Sintonen, S. (2010). *Medialukutaidot osallisuus mediakasvatus* [Media literacies, participation, media education]. Helsinki, Finland: Palmenia/Gaudeamus.

Kupiainen, R., Suoninen, A., & Nikunen, K. (2011). Online habits of Finnish children. Use, risks and data misuse. *Nordicom-Information, 33*(4), 51–57.

Laine, K. (2000). *Koulukuvia. Koulu nuorten kokemustilana* [School images. School as a young people's experience space]. Jyväskylä, Finland: University of Jyväskylä, SoPhi.

Lange, P. (2009). Videos of affinity on YouTube. In. P. Snickars & P. Vonderau (Eds.), *The YouTube reader* (pp. 70–88). Stockholm, Sweden: National Library of Sweden.

Lankshear, C., Gee, J., Knobel, M., & Searle, C. (2002). *Changing literacies.* Philadelphia, PA: Open University Press.

Lankshear, C., & Knobel, M. (2002, February). *Steps toward a pedagogy of tactics.* Paper presented at the National Council of English Teachers' Assembly for Research Mid-Winter Conference, New York, NY. Retrieved from http://everydayliteracies.net/files/pedtact.html

Lankshear, C., & Knobel, M. (2003). *New literacies: Changing knowledge and classroom learning.* Philadelphia, PA: Open University Press.

Lankshear, C., & Knobel, M. (2006). *New literacies: Everyday practices & classroom learning* (2nd ed.). New York, NY: Open Education Press.

Lankshear, C., & Knobel, M. (2007). Sampling "the new" in new literacies. In C. Lankshear & M. Knobel (Eds.), *A new literacies sampler* (pp. 1–24). New York, NY: Peter Lang.

Lankshear, C., & Knobel, M. (2011). *Literacies. Social, cultural and historical perspectives.* New York, NY: Peter Lang.

Lave, J., & Wenger, E. (1991). *Situated learning: Legitimate peripheral participation.* Cambridge, UK: Cambridge University Press.

Leadbeater, C. (2009). *We-Think: Mass innovation, not mass production* (2nd ed.). London, UK: Profile Books.

Leander, K. M. (2007). You won't be needing your laptops today: Wired bodies in the wireless classroom. In C. Lankshear, M. Knobel, C. Bigum, & M. Peters (Eds.), *A new literacies sampler* (pp. 25–48). New York, NY: Peter Lang.

Leander, K. M. (2009). Toward a connective ethnography of online/offline literacy networks. In J. Coiro, M. Knobel, C. Lankshear, & D. Leu (Eds.), *Handbook of research on new literacies* (pp. 33–65). New York, NY: Routledge.

Lefebvre, H. (2007). *The production of space* (D. Nicholson-Smith, Trans.). Oxford, UK: Blackwell.

Lemish, D., & Rice M. L. (1986). Television as a talking book: A prop for language acquisition. *Journal of Child Language, 13*, 251–274.

Lenhart, A. (2012). *Teens, smartphones & texting*. Pew Internet & American Life Project. Retrieved from http://www.pewinternet.org/~/media//Files/Reports/2012/PIP_Teens_Smartphones_and_Texting.pdf

Lenhart, A., Ling, R., Campbell, S., & Purcell, K. (2010). *Teens and mobile phones*. Pew Internet & American Life Project. Retrieved from http://www.pewinternet.org/~/media//Files/Reports/2010/PIP-Teens-and-Mobile-2010-with-topline.pdf

Lenhart, A., Madden, M., MacGill, A., & Smith, A. (2007). *Teens and social media*. Pew Internet & American Life Project. Retrieved from http://www.pewinternet.org/~/media//Files/Reports/2007/PIP_Teens_Social_Media_Final.pdf.pdf

Lessig, L. (2008). *Remix: Making art and commerce thrive in the hybrid economy*. New York, NY: The Penguin Press.

Levy, M., & Michael, R. (2011). Analyzing students' multimodal texts: The product and the process. In M. Thomas (Ed.), *Deconstructing digital natives. Young people, technology and the new literacies* (pp. 83–98). New York, NY: Routledge.

Ling, R. (2004). *The mobile connection: The cell phone's impact on society*. San Francisco, CA: Morgan Kaufmann.

Liu, F. (2011). *Urban youth in China: Modernity, the internet, and the self*. New York, NY: Routledge.

Livingstone, S. (2004). What is media literacy? *Intermedia, 32*(3), 18–20.

Livingstone, S. (2008). Engaging with media—a matter of literacy? *Communication, Culture & Critique, 1*(1) 51–62.

Livingstone, S., Bober, M., & Helsper, E. J. (2005). Active participation or just more information? Young people's take-up of opportunities to act and interact on the Internet. *Information, Communication & Society, 8*(3), 287–314.

Livingstone, S., Haddon, L., Görzig, A., & Ólafsson, K. (2011a). *Risks and safety on the internet. The perspective of European children. Full findings and policy implications from the EU Kids Online Survey of 9–16 year olds and their parents in 25 countries*. Retrieved from http://eprints.lse.ac.uk/33731/

Livingstone, S., Haddon, L. Görzig, A., & Ólafsson, K. (2011b). *EU kids online. Final report*. Retrieved from http://eprints.lse.ac.uk/45490/

Livingstone, S., Ólafsson, K., & Staksrud, E. (2011). *Social networking, age and privacy*. Retrieved from http://eprints.lse.ac.uk/35849/

Lundvall, A. (Ed.). (2009). *Finnish media education policies Approaches in culture and education.* Helsinki, Finland: Helsinki, Finnish Society on Media Education. Retrieved from http://www.mediakasvatus.fi/files/u4/mediaeducationpolicies.pdf

Massey, D. (1994). *Space, place and gender.* Cambridge, UK: Polity Press.

Massey, D. (2010). *For space.* Thousand Oaks, CA: Sage.

McLuhan, M. (1962). *The Gutenberg galaxy: The making of typographic man.* Toronto, Canada: University of Toronto Press.

Mesch, G., & Talmud, I. (2010). *Wired youth. The social world of adolescence in the information age* [Kindle iPad version]. New York, NY: Routledge. Retrieved from http://www.amazon.com

Messaris, P. (1993). *Visual literacy. Image, mind, & reality.* Boulder, CO: Westview Press.

Meyrowitz, J. (1985). *No sense of place: The impact of electronic media on social behaviour.* New York, NY: Oxford University Press.

Mikkola, A.-M., Luukka, M.-R., & Ahonen, K. (2006). *Voima. Äidinkieli ja kirjallisuus 8* [Power. Mother tongue and literature 8]. Helsinki, Finland: WSOY.

Mikkola, A.-M., Luukka, M.-R., & Ahonen, K. (2005). *Taito. Äidinkieli ja kirjallisuus 7* [Skill. Mother tongue and literature 7]. Helsinki, Finland: WSOY.

Moje, E. B., Mcintosh Ciechanowski, K., Kramer, K., Lindsay, E., Carrillo, R., & Collazo, T. (2004). Working toward third space in content area literacy: An examination of everyday funds of knowledge and discourse. *Reading Research Quarterly, 39*(1), 38–70.

Moores, S. (2012). *Media, place & mobility.* New York, NY: Palgrave Macmillan.

Morley, D. (2000). *Home territories: Media, mobility and identity.* New York, NY: Routledge.

Myllyniemi, S. (2009). *Taidekohtia, Nuorisobarometri 2009* [Art-points. Youth barometer 2009]. Jyväskylä, Finland: Jyväskylä, Finnish Youth Research Network/Finnish Youth Research Society, publications 97, Nuorisoasian neuvottelukunta, julkaisuja 41. Retrieved from http://www.minedu.fi/OPM/Nuoriso/nuorisoasiain_neuvottelukunta/julkaisut/barometrit/liitteet/Nuorisobarometri_2009.pdf

Negroponte, N. (1996). *Being digital.* New York, NY: Vintage.

Nespor, J. (1997). *Tangled up in school: Politics, space, bodies, and signs in the educational process.* Mahwah, NJ: Lawrence Erlbaum.

Niemi, E. K. (2012). Opettajakysely [Teacher survey]. In E. Niemi & E. K. (Eds.), *Aihekokonaisuuksien tavoitteiden toteutumisen seuranta-arviointi 2010* [Assessment of the goals of cross-curricular themes 2010] (Education monitoring report 2012: 1) (pp. 18–41). Helsinki, Finland: Helsinki, Finnish National Board of Education.

Niemi, H. (2012). The societal factors contributing to education and schooling in Finland. In H. Nemi, A. Toom, & A. Kallioniemi (Eds.), *Miracle of education. The principles and practices of teaching and learning in Finnish schools* (pp. 17–38). Rotterdam, the Netherlands: Sense Publishers.

Niemi, H., Kynäslahti, H., & Vahtivuori-Hänninen, S. (2012). Towards ICT in everyday life in Finnish schools: Seeking conditions for good practices. *Learning, Media and Technology,* doi: 10.1080/17439884.2011.651473

Nikunen, K. (2010). Online among us. Experiences of virtuality in everyday life. *Nordicom-Information, 32*(2–3), 75–82.

OECD. (2006). *Equity in education: Thematic review.* Retrieved from http://www.oecd.org/dataoecd/49/40/36376641.pdf

OECD. (2010). *PISA 2009 results: What students know and can do: Student performance in reading, mathemetics and science* (Volume 1), doi: 10.1787/9789264091450-en

Ofcom. (2006). *Media literacy audit–report on adult media literacy.* Retrieved from http://stakeholders.ofcom.org.uk/market-data-research/media-literacy/archive/medlitpub/medlitpubrss/medialit_audit/

Ólafsson, K. (2011). Nordic children's risks and opportunities online. The EU Kids Online survey from a Nordic perspective. *Nordic Information, 33*(4), 37–50.

Ong, W. (1994). *Secondary orality and secondary visualism.* Retrieved from http://cdm.slu.edu/cdm/singleitem/collection/ong/id/668/rec/18

Opetushallitus. (2011). Demokratiaselvitys [Democracy report]. (Reports of Finnish National Board of Education, 2011: 27). Helsinki, Finland: Helsinki, Finnish National Board of Education.

Pahl, K., & Rowsell, J. (2005). *Literacy and education: Understanding the new literacy studies in the classroom.* London, UK: Paul Chapman.

Palmu, T. (2003). *Sukupuolen rakentuminen koulun kulttuurisissa teksteissa. Etnografia yläkoulun äidinkielen oppitunneilla* [Construction of gender in cultural texts in school. An ethnographic research on mother tongue lessons in secondary school] (Doctoral dissertation). University of Helsinki, Department of Education. Retrieved from https://helda.helsinki.fi/handle/10138/19881

Papacharissi, Z. A. (2010). *A private sphere: Democracy in a digital age.* Cambridge, UK: Polity Press.

Pearlman, R. (2009). Making twenty-first century schools: Creating learner-centered schoolplaces/workplaces for a new culture of students at work. *Educational Technology, 49*(5), 14–19.

Prensky, M. (2001). Digital natives, digital immigrants. *On the Horizon, 9*(5), 1–6.

Rajanti, T. (2008). Wormholes as a new spatial paradigm: Social production of space in postfordist society and the art of studying it. *Sociologisk Årbok*,1, 76–84.

Rancière, J. (1991). *The ignorant schoolmaster: Five lessons in intellectual emancipation* (K. Ross, Trans.). Palo Alto, CA: Stanford University Press.

Rancière, J. (2004, December). Introducing disagreement (S. Corcoran, Trans.). *Angelaki, Journal of Theoretical Humanities, 9*(3),3–9.

Rancière, J. (2009). *The emancipated spectator* (G. Elliott, Trans.). New York, NY: Verso.

Rantala, L., & Suoranta, J. (2009). Digital literacy policies in the EU—inclusive partnership as the final stage of governmentality. In C. Lankshear & M. Knobel (Eds.), *Digital literacies: Concepts, policies and practices* (pp. 91–117). New York, NY: Peter Lang.

Rodriques, M. J. (2006). The European way to a knowledge-intensive-economy—the Lisbon strategy. In M. Castells & G. Cardoso (Eds.), *The network society: From knowledge to policy* (pp. 405–424). Washington, DC: Johns Hopkins Center for Transatlantic Relations.

Salomon, G. (1990). The computer lab: A bad idea now sanctified. *Educational Technology*, 30(10), 50–52.

Sanomalehtien liitto. (2012, February 1). *Sanomalehteä aktiivisesti lukevat nuoret pärjäsivät PISA:ssa* [Active young newspaper readers had good results in PISA test]. Retrieved from http://www.sanomalehdet.fi/index.phtml?s=2787

Schulz, W., Ainley, J., Fraillon, J., Kerr, D., & Losito, B. (2010). *Initial findings from the IEA International Civic and Citizenship Education Study*. Retrieved from http://iccs.acer.edu.au/uploads/File/Reports/ICCS_10_Initial_Findings.pdf

Sesame Street. (1972). Ernie eats a banana and gives Bert the peel [Video file]. Retrieved from http://www.youtube.com/watch?v=IAaSx2zv8ow

Selwyn, N. (2011). *Schools and schooling in the digital age: A critical analysis*. New York, NY: Routledge.

Sennett, R. (2009). *The craftsman*. London, UK: Penguin Books.

Share, J. (2009). *Media literacy is elementary. Teaching youth to critically read and create media*. New York, NY: Peter Lang.

Sobchack, V. (1992). *The address of the eye. A phenomenology of film experience*. Oxford, UK: Princeton University Press.

Soep, E., & Chávez, V. (2010). *Drop that knowledge: youth radio stories* [Kindle iPad version]. Berkeley, University of California Press. Retrieved from http://www.amazon.com

Soja, E. W. (1996). *Thirdspace: Journeys to Los Angeles and other real-and-imagined places*. Cambridge, UK: Blackwell.

Sonck, N., Kuiper, E., & de Haan, J. (2012). Digital skills in the context of media literacy. In S. Livingstone, L. Haddon, & A. Görtig (Eds.), *Children, risk and safety on the internet. Research and policy challenges in comparative perspective* (pp. 87–98). Bristol, UK: The Policy Press.

Stald, G. (2008). Mobile identity: Youth, identity, and mobile communication media. In D. Buckingham (Ed.), *Youth, identity, and digital media* (pp. 143–164). Cambridge, MA: The MIT Press.

Suoninen, A., Kupari, P., & Törmäkangas, K. (2010). *Nuorten yhteiskunnalliset tiedot, osallistuminen ja asenteet. Kansainvälisen ICCS–tutkimuksen ensituloksia* [Civic skills, participation and attitudes of the youth. First results from the international ICCS study]. Retrieved from http://ktl.jyu.fi/img/portal/17876/ICCS_2009_verkkoversio_v2.pdf

Suoranta, J., & Vadén, T. (2010). *Wikiworld* [e-book]. Retrieved from http://wikiworld.wordpress.com/

Säljö, R. (2001). *Oppimiskäytännöt. Sosiokulttuurinen näkökulma* [Learning practices. Sociocultural perspective] (B. Grönholm, Trans.). Helsinki, Finland: WSOY.

Taalas, P., Tarnanen, M., Kauppinen, M., & Pöyhönen, S. (2008). Media landscapes in school and in free time: Two parallel realities? *Digital Kompetanse*, 3(4), 240–256.

Takahashi, T. (2011). Japanese youth and mobile media. In M. Thomas (Ed.), *Deconstructing digital natives. Young people, technology and the new literacies* (pp. 67-82). New York, NY: Routledge.

Tapscott, D. (1999). *Growing up digital: The rise of the net generation* [Kindle iPad version]. New York, NY: McGraw-Hill. Retrieved from http://www.amazon.com

Tapscott, D., & Williams, A. D. (2006). *Wikinomics: How mass collaboration changes everything*. New York, NY: Portfolio.

Thomas, A. (2007). *Youth online: Identity and literacy in the digital age*. New York, NY: Peter Lang.

Thomas, D., & Brown, J. S. (2011). *A new culture of learning. Cultivating the imagination for a world of constant change* [Kindle iPad version]. Retrieved from http://www.amazon.com

Thomas, M. (2011a). *Deconstructing digital natives. Young people, technology and the new literacies*. New York, NY: Routledge.

Thomas, M. (2011b). Technology, education, and the discourse of the digital native. In M. Thomas (Ed.), *Deconstructing digital natives. Young people, technology and the new literacies* (pp. 1–11). New York, NY: Routledge.

Toivanen, T. (2012). Drama education in the Finnish school system: Past, present and future. In H. Nemi, A. Toom, & A. Kallioniemi (Eds.), *Miracle of education. The principles and practices of teaching and learning in Finnish schools* (pp. 227–236). Rotterdam, the Netherlands: Sense Publishers.

Torney-Purta, J., Lehmann, R., Oswald, H., & Schulz, W. (2001). *Citizenship and education in twenty-eight countries: Civic knowledge and engagement at age fourteen*. Retrieved from http://www.terpconnect.umd.edu/~jtpurta/interreport.htm

Turkle, S. (2011). *Alone together: Why we expect more from technology and less from each other* [Kindle iPad version]. New York, NY: Basic Books. Retrieved from http://www.amazon.com

Turow, J. (2011). *The daily you. How the new advertising industry is defining your identity and your worth* [Kindle iPad version]. New Haven, CT: Yale University Press. Retrieved from http://www.amazon.com

Tyack, D., & Tobin, W. (1994). The "grammar" of schooling: Why has it been so hard to change? *American Educational Research Journal, 31*(3), 453–479.

Tyner, K. R. (1998). *Literacy in a digital world: Teaching and learning in the age of information*. Mahwah, NJ: Lawrence Erlbaum.

Uusitalo, N., Vehmas, S., & Kupiainen, R. (2011). *Naamatusten verkossa. Lasten ja nuorten mediaympäristön muutos, osa 2* [Facing one another online: The changing media environment of children and young people]. Retrieved from http://urn.fi/urn:isbn:978-951-44-8354-7

Vahtivuori-Hänninen, S., & Kynäslahti, H. (2012). ICTs in school's everyday life. In H. Nemi, A. Toom, & A. Kallioniemi (Eds.), *Miracle of education. The principles and practices of teaching and learning in Finnish schools* (pp. 237–248). Rotterdam, the Netherlands: Sense Publishers.

Virta, A., & Yli-Panula, E. (2012). History, social science and geography education. In H. Nemi, A. Toom, & A. Kallioniemi (Eds.), *Miracle of education. The principles and practices of teaching and learning in Finnish schools* (pp. 189–207). Rotterdam, the Netherlands: Sense Publishers.

Virta, T. (1994). *Saippuaoopperat ja suomalaiset naiset. [Soap operas and female Finnish viewers]*. Publications 43. Jyväskylä: Finland: University of Jyväskylä, Research Centre for Contemporary Culture.

Wells, P. (1998). *Understanding animation*. New York, NY: Routledge.

Wenger, E. (2006). *Communities of practice. A brief introduction*. Retrieved from http://www.ewenger.com/theory/

Wenger, E. (2008). *Communities of practice. Learning, meaning and identity*. New York, NY: Cambridge University Press.

Willett, R. (2009). Young people's video productions as new sites of learning. In V. Carrington & M. Robinson (Eds.), *Digital literacies: Social learning and classroom practices* (pp. 13–26). Thousand Oaks, CA: Sage.

Williamson, J. (1981). How does girl number twenty understand ideology? *Screen Education, 40*(2), 80–87.

Xu Fan (2007, July 26). Self organized citizen translations of Harry Potter. Retrieved from http://www.zonaeuropa.com/20070728_1.htm

INDEX

new literacies

AND DIGITAL EPISTEMOLOGIES

Colin Lankshear & Michele Knobel

General Editors

New literacies emerge and evolve apace as people from all walks of life engage with new technologies, shifting values and institutional change, and increasingly assume 'postmodern' orientations toward their everyday worlds. Despite many efforts to take account of such changes, educational institutions largely remain out of touch with the range of new ways of making and sharing meanings that increasingly mediate and shape the lives of the young people they teach and the futures they face. This series aims to explore some key dimensions of the changes occurring within social practices of literacy and the educational challenges they present, with a view to informing educational practice in helpful ways. It asks what are new literacies, how do they impact on life in schools, homes, communities, workplaces, sites of leisure, and other key settings of human cultural engagement, and what significance do new literacies have for how people learn and how they understand and construct knowledge. It aims to challenge established and 'official' ways of framing literacy, and to ask what it means for literacies to be powerful, effective, and enabling under current and foreseeable conditions. Collectively, the works in this series will help to reorient literacy debates and literacy education agendas.

For further information about the series and submitting manuscripts, please contact:

Michele Knobel & Colin Lankshear
Montclair State University
Dept. of Education and Human Services
3173 University Hall
Montclair, NJ 07043
michele@coatepec.net